MATLAB® in bioscience and biotechnology

Woodhead Publishing Series in Biomedicine

Published by Woodhead Publishing Limited

Published by Woodhead Publishing Limited

Published by Woodhead Publishing Limited

Published by Woodhead Publishing Limited

Published by Woodhead Publishing Limited

Woodhead Publishing Series in Biomedicine: Number 3

MATLAB® in bioscience and biotechnology

Leonid Burstein

WP

WOODHEAD
PUBLISHING

Oxford   Cambridge   Philadelphia   New Delhi

Published by Woodhead Publishing Limited

Woodhead Publishing Limited, 80 High Street, Sawston, Cambridge, CB22 3HJ, UK
www.woodheadpublishing.com
www.woodheadpublishingonline.com

Woodhead Publishing, 1518 Walnut Street, Suite 1100, Philadelphia, PA 19102-3406, USA

Woodhead Publishing India Private Limited, G-2, Vardaan House, 7/28 Ansari Road,
Daryaganj, New Delhi – 110002, India
www.woodheadpublishingindia.com

First published in 2011 by Biohealthcare Publishing (Oxford) Limited; republished in 2012 by Woodhead
Publishing Limited
ISBN: 978-1-907568-04-6 (print) and ISBN: 978-1-908818-03-4 (online)
Woodhead Publishing Series in Biomedicine ISSN 2050-0289 (print); ISSN 2050-2097 (online)

MATLAB, Bioinformatics toolbox and Simulink are registered trademarks of The MathWorks, Inc.

Typeset by Domex e-Data Pvt. Ltd., India
Printed in the UK and USA

Published by Woodhead Publishing Limited

In memory of my father Matvey.
To my mother Leda, my wife Inna, and my son Dmitri

Published by Woodhead Publishing Limited

Contents

Published by Woodhead Publishing Limited

Preface

In the last few decades, two seemingly disparate sciences – computer science and biology – have interpenetrated and affected one another. We see students enrolled in computer science beginning their careers in biotechnological laboratories, and biotechnologists creating bio-computers and being active in computer science. There appears to be an urgent need to familiarize biotechnologists with the same computing tools as are imparted to technicians.

This book represents a short introduction to MATLAB® oriented towards various collaborative areas of biotechnology and bioscience. My hope is that it will be equally useful to undergraduate and graduate students and to practising engineers. It concentrates on the fundamentals of MATLAB® and gives examples of its application to a wide range of current bioengineering problems in computational biology, molecular biology, biokinetics, biomedicine, bioinformatics and biotechnology. In the last decade MATLAB® has been presented to students as a basic computational tool that they need to learn. Consequently, many students unfamiliar with programming, engineers and scientists have come to regard it as user-friendly and highly convenient in solving their specific problems. Numerous books are available on programming in MATLAB® for engineers in general, irrespective of their specialization, or for those specializing in some specific area, but none has been designed specifically for a wide, interdisciplinary, topical area such as bioengineering. Thus, MATLAB® is presented here with examples and applications to various school- and advanced bioengineering problems – from growing populations of microorganisms and population dynamics, to reaction kinetics and reagent concentrations, predator–prey models, mass-transfer problems, and to sequence analysis and sequence statistics.

The book distills my experience of many years of MATLAB® teaching in introductory and advanced courses for students, engineers and scientists specializing in bioscience and engineering.

I would like to thank the people who attracted me to the subject and thereby played key roles in the inception and appearance of this book: my colleague Professor Rosa Azhari (Biotechnology Department, ORT Braude College), and software support team head Moshe Barak (Computer Center, Technion – Israel Institute of Technology). I also thank MathWorks Inc. (3 Apple Hill Drive, Natick, MA 01760-2098, USA, Tel: 508-647-7000, Fax: 508-647-7001, E-mail: info@mathworks.com, Web: www.mathworks.com) who graciously granted permission to reproduce material appearing in this book.

I thank Ing. Eliezer Goldberg, former resident scientific editor at Technion, for patience and invaluable editorial assistance, and would also like to thank Dr Glyn Jones, head of Biohealthcare Publishing (Oxford) Ltd, for invaluable support throughout all stages of publication of this book.

I hope this book will prove useful to students and engineers in both natural and life sciences and provide them with an opportunity to work with one of the finest software tools.

Any reports of errata or bugs, comments and suggestions on the book's contents will be accepted gratefully by the author.

Leonid Burstein
Nesher, Haifa, Karmiel, Israel
September 2010

List of figures and tables

Figures

Published by Woodhead Publishing Limited

Published by Woodhead Publishing Limited

Tables

Published by Woodhead Publishing Limited

Published by Woodhead Publishing Limited

About the author

Leonid Burstein is Senior Lecturer at large at Technion – Israel Institute of Technology, at the ORT Braude College, in the Biotechnology and Software Engineering Departments, and at a number of other universities and highschools in Western and Lower Galilee.

Following an MA in thermophysics at Lomonosov Technological Institute at Odessa, Ukraine, and a doctorate at the National Research Institute for Physical and Radio Engineering Measurements at Moscow, he obtained his PhD in physical properties of materials from the Heat/Mass Transfer Institute of the Belarus Academy of Science, Minsk, in 1974. After a short period of work in Russia and Belarus, Dr Burstein started his carrier at the Piston Ring Institute in Odessa, where he served from 1974 to 1990 as Head of Projects and Head of the CAD/CAM group. In 1991, he began work at the Technion – IIT, Israel, at the Faculty of Mechanical Engineering, in the Quality Assurance and Reliability Program at the Faculty of Industrial Engineering and Management, and at the Taub Computer Centre as an advisor on MATLAB® and other scientific software. He also worked at the Technion Research and Development Foundation as principal researcher in funded projects in various areas such as diesel tribology and environment control. He also taught various courses at Haifa University, at the Technion, at the Kinnereth Academic College and elsewhere. He currently teaches a MATLAB® course for biotechnologists at ORT Braude College.

He is an Editorial Board Member and reviewer for a number of international journals and a Committee Member of numerous conferences. He is the author of several patents, has published four chapters in scientific books and authored/co-authored more than 60 publications in leading scientific journals.

Published by Woodhead Publishing Limited

He can be contacted at:

Technion – Israel Institute of Technology
Technion City
32000 Haifa
Israel
E-mail: *leonidburstein@gmail.com* (prefereble) or *leonidb@technion.ac.il*

Published by Woodhead Publishing Limited

1

Introduction

Everything that can be counted – should be.
Anonymous

Biological engineering is defined as application of engineering principles to the widest spectrum of living systems – from molecular biology, biochemistry and microbiology, to bio-medicine, genetics and bioinformatics. And as in general engineering, computers and the ability to use them are vitally important. This is true also for other professionals of any bio-industry. Thus, bio-specialists and scientists working in these areas need to have the computational resources to be able to solve various problems. A widespread and powerful tool for such purposes is MATLAB® – the software for technical computing. It is designed to solve both general and specific problems; of these, the latter are treated with so-called toolboxes, which currently include means specialized for bio-problems. An obstacle to the effective understanding and implementation of MATLAB® in practice is the inadequate level of math reached by students and specialists in areas of bioscience, combined with a lack of textbooks tailored to such audiences. This book is intended as a remedy. It is organized as follows.

I begin by covering primary MATLAB® programming and then move to more complicated problems by means of this language; the material is illustrated throughout by examples from different areas of bioengineering and biological science. The topics were chosen on the basis of several years of teaching MATLAB® for biotechnologists and they are presented so that inexperienced users can progress gradually, with the previously presented material being the only prerequisite for each new chapter.

Chapter 2 introduces the MATLAB® environment, language design, help options, variables, matrix and array manipulations, elementary and special functions, flow chart control, conditional statements and other basic MATLAB® features.

Published by Woodhead Publishing Limited

In chapter 3 the plotting tool is described by using examples of graphic presentation in various calculations. Mastering the material in chapters 2 and 3 will allow readers to create their own MATLAB® programs.

Chapter 4 presents the MATLAB® script- and m-files; the commands for numerical integration, differentiation, inter-/extrapolation and curve fitting, together with their various applications, are given.

In chapter 5, particular solutions for ordinary and partial differential equations are briefly presented together with examples from bio-systems involving a single differential equation or a set. This chapter assumes a somewhat greater familiarity with mathematics.

In the final chapter the bioinformatics tool is introduced through applications employed in sequence analysis and statistics. Emphasis is placed on DNA and protein sequence database access and further pairwise or multiple alignments.

The Appendix details the studied MATLAB® commands and functions.

Application problems included at the end (and sometimes in the middle) of each chapter are solved with commands accessible to the reader; the solutions are not necessarily the shortest or most original, but should be easy to understand and follow up. Readers are invited to write their own solutions and check the results against those given herein. At the end of each chapter are questions and problems, and readers are encouraged to attempt them for better assimilation of the material. The contexts and values used in the problems are not factual and are intended for learning purposes only.

The MATLAB® used in the book is R2010a, version 7.10.0. Each subsequent version should incorporate all previous ones; hence, the fundamental commands given here should be valid in future versions. It is assumed that the user has a computer with MATLAB® installed on it and is able to perform basic computer operations.

Each command is explained here in its simplest form; additional information is available in the MATLAB®-help or original MATLAB® documentation.

Let us begin.

2

MATLAB® basics

MATLAB® came into being in the 1970s as a tool for mathematicians and educators, but was soon adopted by engineers as an effective means for technical computing. Its name is a composite of the words 'Matrix' and 'Laboratory', emphasizing that its main element is the matrix. Such an approach permitted unification of the processes of various calculations, graphics, modeling, simulation and algorithm development. This chapter introduces the main windows and starting procedure, describes the main commands for simple arithmetic, algebraic and matrix operations, and presents the basic loops and relational and logical operators.

2.1 Starting with MATLAB®

MATLAB® can be installed on computers running different operation systems, but I will assume here that the reader uses a personal computer running a Windows operating system. To start one has simply to click on the MATLAB® icon (Figure 2.1) provided with a MATLAB® subscription; the icon is placed on the Quick Lunch bar or on the Windows Desktop. Another way to start the program is to select MATLAB® 20010a in the MATLAB®-directory in the 'All Programs' option of the Windows 'Start' menu.

2.1.1 MATLAB® Desktop and its windows

The window that first opens is the MATLAB® Desktop (Figure 2.2), which comprises four windows: Command, Current Folder, Workspace and Command History.

These are the most intensively used windows and are briefly described further. There are also Help, Editor and Figure windows that do not appear

Published by Woodhead Publishing Limited

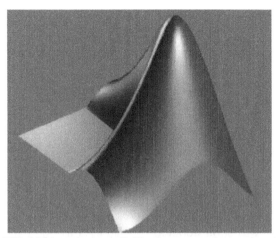

Figure 2.1   MATLAB® icon (enlarged). The image can be produced with the `logo` command; the background color has been changed.

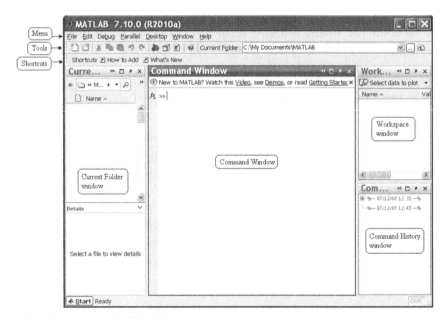

Figure 2.2   MATLAB® desktop.

Published by Woodhead Publishing Limited

with the MATLAB® Desktop and are described in the chapters where they are used.

The Desktop also contains: the Menu, which can be changed depending on the tool being used; the MATLAB® Tools bar, which contains the more common functions; the Shortcuts bar, where one can place icons for quick running of MATLAB® programs or group commands; and the Start button, used to access various tools, demos, shortcuts and documentation.

The **Command Window** is the main outlet where commands are entered and results are displayed. Sometimes it is convenient to separate it from the desktop by clicking ⏴ to the right of the title bar. Such separation is possible for all Desktop windows. To combine windows one has to click on ⏴ or select the Default line in the Desktop Layout of the Desktop option at the Menu bar.

Workspace is the graphical interface that allows us to view and manage the variables and other objects of the MATLAB® workspace; it also displays and automatically updates the values of each variable.

Current Folder presents a browser that shows the full path to the current folder, and shows the contents of the current folder. When starting MATLAB®, we view a starting directory which is called the startup directory. After selecting the file, information about it appears in the Details panel.

Command History stores the commands most recently entered in the Command Window.

2.1.2 Elementary functions and interactive calculations

Two main working modes are available in MATLAB® – interactive and with m-files. I will explain the latter in later chapters. The interactive mode is discussed briefly here.

To enter and execute a command, it must be typed in the Command Window immediately after the command prompt >>. Figure 2.3 shows this window with some elementary commands.

The symbol f_x, which appears in the most recent versions of MATLAB®, is called the Function Browser, and helps to find the function required and information about its syntax and usage.

Entering a command and manipulating with it require us to master the following operations:

- the command must be typed next to the prompt >>;
- the Enter key must be pressed for execution;

Published by Woodhead Publishing Limited

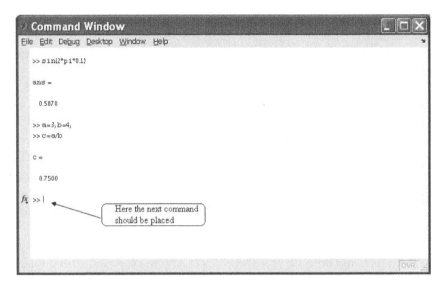

Figure 2.3   Command Window; the view after separation from the desktop.

- a command in a preceding line cannot be changed; to correct or repeat an executed command the up-arrow key ↑ should be pressed;
- a long command can be continued in the next line by typing … – three periods; commands in the same line should be divided by semicolons (;) or by commas (,); a semicolon at the end of a command prevents the answer from being displayed;
- the symbol % (percentage symbol) designates those comments that should be written after it in the line, and the comments are not executed after entering;
- the command clc clears the Command Window.

The Command Window can be used as a calculator by using the following symbols for arithmetical operations: + (addition), − (subtraction), * (multiplication), / (right division), \ (left division, used mostly for matrices), ^ (exponential function).

These operations are applicable to a wide variety of elementary and trigonometric functions that should be written as the name with the argument in parentheses, e.g. sin x should be written as sin(x); in trigonometric functions the argument x should be given in radians. A short list of such functions and variables is given in Table 2.1. Hereinafter the operations executed in the Command Window are written after the command line prompt (>>), and the user will need to press the Enter key after entering one or more commands written in one command line.

Published by Woodhead Publishing Limited

Table 2.1 Elementary and trigonometric mathematical functions

Functions and constants in Math	MATLAB® presentation	MATLAB® example (inputs and outputs)
$\lvert x \rvert$ – absolute value	abs(x)	>> abs(−15.1234) ans = 15.1234
e^x – exponential function	exp(x)	>> exp(2.7) ans = 14.8797
$\ln x$ – natural (base e) logarithhm	log(x)	>> log(10) ans = 2.3026
$\log x$ – Napierian (base 10) logarithm	log10(x)	>> log10(10) ans = 1
$\sqrt{x}$ – square root	sqrt(x)	>> sqrt(2/3) ans = 0.8165
π – the number π	pi	>> 2*pi ans = 6.2832
Round towards minus infinity	floor(x)	>> floor(−12.1) ans = −13
Round to the nearest integer	round(x)	>>round(12.6) ans = 13
$\sin x$ – sine	sin(x)	>> sin(pi/3) ans = 0.8660
$\cos x$ – cosine	cos(x)	>> cos(pi/3) ans = 0.5000
$\tan x$ – tangent	tan(x)	>> tan(pi/3) ans = 1.7321
$\cot x$ – cotangent	cot(x)	>> cot(pi/3) ans = 0.5774
$\arcsin x$ – inverse sine	asin(x)	>> asin(1) ans = 1.5708
$\arccos x$ – inverse cosine	acos(x)	>> acos(1) ans = 0
$\arctan x$ – inverse tangent	atan(x)	>> atan(1) ans = 0.7854
$\operatorname{arccot} x$ – inverse cotangent	acot(x)	>> acot(1) ans = 0.7854
$n!$ – factorial	factorial(n)	>> factorial(5) ans = 120

Published by Woodhead Publishing Limited

The result of entering a command is a variable with name `ans`. The equal sign (=) is called the assignment operator and is used to specify a value to a variable, e.g. to the `ans`. An entered new value cancels its predecessor.

Arithmetic operations are performed in the following order: operations in parentheses (starting with the innermost), exponentiation, multiplication and division, addition and subtraction. If an expression contains operations of the same priority, they run from left to right.

Examples of arithmetic operations in the Command Window are given below:

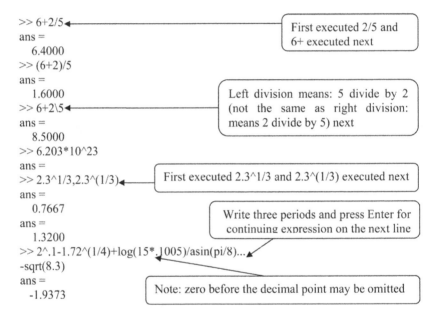

The outputted numbers are displayed here in `short` format (default format) – a fixed point followed by four decimal points. The format can be changed to `long`, 14 digits after the point, by typing the command: `format long`. To return to the default format the user has to type `format`.

There are other formats that can be obtained by typing `help format`; the word after `help` appears in blue, for ease of viewing.

2.1.3 Help and Help Window

For information about use of some commands, type and enter `help` with the command name after a space next to this word, e.g. `help format` as above. The explanations appear immediately after this in the Command Window.

Published by Woodhead Publishing Limited

For a command concerning a particular topic of interest, the lookfor command may be used. For example, for the name of MATLAB® command(s) on the subject of codons one should enter lookfor codon and the commands will subsequently appear on the screen, as shown below:

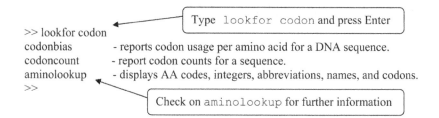

>> lookfor codon
codonbias   - reports codon usage per amino acid for a DNA sequence.
codoncount   - report codon counts for a sequence.
aminolookup   - displays AA codes, integers, abbreviations, names, and codons.
>>

Type lookfor codon and press Enter

Check on aminolookup for further information

For further information the user has to click on the selected command or again use the help command. To interrupt the search process, the two abort keys Ctrl and c should be clicked together; these keys should also be used to interrupt any other process, e.g. that of program/command execution.

For more detailed information one can similarly use the doc command, e.g. doc aminolookup, in which case the Help window will be opened. The Help window can also be opened by selecting the Product Help line in the Help options on the MATLAB® Desktop menu line (Figure 2.4).

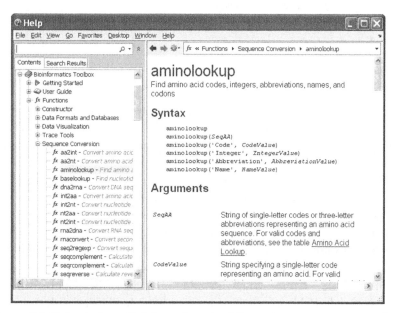

Figure 2.4 Help window with information about the aminolookup command.

Published by Woodhead Publishing Limited

The Help window comprises three panes: on the left are the Contents or Search Results and on the right is the page containing information on the topic. Information on any subject is obtainable by typing the word(s) into the search line in the upper left-hand corner. The Search Results pane shows a preview of where the search words were found within the page, and the concrete information is displayed on the right.

2.1.4 Variables and commands for management of variables

A variable is a symbolic term written as a letter(s) and associated with a concrete numerical value. MATLAB® allocates memory space for storage of variable names and their values. A variable can be a scalar – a single number – or an array – a table of numbers. The name can be as many as 63 characters long, and contain letters, digits and underscores, but the first character must be a letter. Existing commands (sin, cos, sqrt, etc.) cannot be used as names.

The assignment and usage of variables in algebraic calculations is demonstrated next.

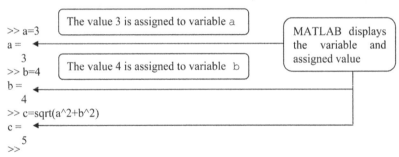

Predefined MATLAB® variables can be used without being assigned. Except for the previously mentioned pi and ans, these are inf (infinity), i or j (square root of –1), and NaN (not-a-number, used when a numerical value is moot, e.g. 0/0).

The following commands can be used for management of variables: clear, to remove from memory; clear x y, for removing named variables x and y only; who for displaying the names of variables; or whos for displaying variable names, matrix sizes, variable byte sizes and variable classes. This information can also be obtained in the Workspace Window, where each variable is presented by the icon ⊞ with the same information as in the case of whos but with additional data; the popup menu for selection of desirable information appears by right-clicking with the cursor placed on the Workspace Window menu line.

Published by Woodhead Publishing Limited

2.1.5 Output commands

As previously noted, MATLAB® automatically displays the result after each command is entered, but does not display it if the command is followed by a semicolon. MATLAB® has additional display commands, the two most frequently used of which are `disp` and `fprintf`.

The `disp` command is used to display text or variable values without the name of the variable and the equal sign. Each new `disp` command yields its result in a new line. In general form the command reads

> `disp('Text string')` or `disp(Variable name)`

The text between quotes is displayed in blue.

For example:

```
>> Na=6.0221*10^23
Na =
   6.0221e+023
>> disp('Avogadro Const'),disp(Na)
Avogadro Const
   6.0221e+023
```

Display the variable value without the `disp` command

The first `disp` command used to display string `Avogadro Const` above the Avogadro's number value displayed by the second `disp` command

The `fprintf` command is used to display text and data or to save them to file. The command has various forms that present difficulties for beginners, and here I give the simplest of them for displaying the results of a calculation.

To display text and a number on the same line the following form is used:

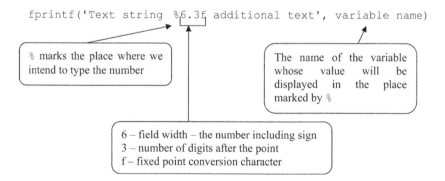

`fprintf('Text string %6.3f additional text', variable name)`

`%` marks the place where we intend to type the number

The name of the variable whose value will be displayed in the place marked by `%`

6 – field width – the number including sign
3 – number of digits after the point
f – fixed point conversion character

To divide a text into two or more lines, or starting with a new line, \n (slash *n*) must be written before the word or sign that we want to see on the new line.

The field width and the number of digits after the point (6.3 in the example presented) are optional; the sign % and the character f, called conversion character, are obligatory. The character f specifies the fixed point notation in which the number is displayed. Some additional notations that can be used are: i, integer, e, exponential (e.g. 2.309123e+001); and g, the more compact form of e or f, with no trailing zeros.

Addition of several %f units (or full formatting elements) permits inclusion of multiple variable values in the text. For example, using the fprintf command:

>> m_H2=2.01588; m_CL2 =70.906;
>> fprintf(' Mass of H_2 is %6.3f g/mol\n Mass of Cl_2 is %6.3f g/mol\n',m_H2,m_CL2)
 Mass of H_2 is 2.016 g/mol
 Mass of Cl_2 is 70.906 g/mol
>>

The color of the text in quotes is the same as in disp (blue).

The commands described can be used to output tables as will be shown later, after introduction of vectors and matrices.

2.1.6   Application examples

2.1.6.1   DNA volume

As shown here

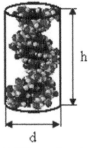

the idealized volume of the DNA molecule can be calculated using the expression for the volume of a cylinder:

$$V = \pi (d/2)^2 h$$

where r, the radius of the DNA molecule, is about 1.58×10^{-3} μm, and h, its length, is 3.34×10^{-3} μm.

Problem: Calculate the volume of the DNA molecule.
  The solution:

```
>> d=0.00158;
>> h=0.000334;
>> V_DNA=pi*(d/2)^2*h
V_DNA =
  6.5486e-010
>>
```

Assign values to the variables d and h

Calculate the DNA volume

2.1.6.2 The distance between two molecules

The distance d between two molecules shown in a figure in a Cartesian coordinate system is given by the expression

$$d = \sqrt{(x_1 - x_2)^2 + (y_1 - y_2)^2 + (z_1 - z_2)^2}$$

where x, y and z are the coordinates, and subscripts 1 and 2 denote the first and second molecules, respectively. The dimensionless coordinates are: $x_1 = 0.1$, $y_1 = 0.02$, $z_1 = 0.12$, $x_2 = 0.2$, $y_2 = 0.5$, $z_2 = 0.11$.

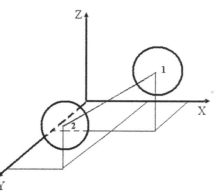

Problem: Calculate the distance for the given coordinates of the molecules.
  The solution:

Assign coordinate values to the variables x1, y1, z1, x2, y2, z2

```
>> x1=0.1;y1=0.02;z1=0.12;
>> x2=0.2;y2=0.5;z2=0.11;

>> d=sqrt((x2-x1)^2+(y2-y1)^2+(z2-z1)^2)
d =
  0.4904
>>
```

Calculate the distance

2.1.6.3 Cooling of a fluid according to Newton's law

The time t taken for a liquid (e.g. coffee in a cup) to cool from temperature T_0 to T is given from Newton's law, according to

$$t = -\frac{1}{k}\ln\frac{T-T_s}{T_0-T_s}$$

where T_s is the ambient temperature, T_0 is the initial temperature and k is a constant. If the coffee has an initial temperature of 70 °C, the ambient temperature is 20 °C and k is 0.3 C/min.

Problem: When will the coffee be fit to drink ($T = 28$ °C)?
    The solution:

```
>> T₀=70;
>> Tₛ=20;
>> T=28;
>> k=0.3;
>> t=-1/k*log((T - Tₛ)/(T₀ - Tₛ))
t =
 6.1086
>>
```

2.1.6.4 Constant of chemical reaction

The rate constant k of a chemical reaction is given by the Arrhenius equation:

$$k = A\,e^{\frac{-E_a}{RT}}$$

where E_a is the activation energy, A is the frequency of molecular collision, T is the temperature at which the reaction passes and R is the gas constant. If for a dissociation reaction these parameters are $E_a = 75{,}000$ J/mol, $A = 1 \times 10^{14}$ s^{-1}, $T = 300$ K and $R = 8.314$ J/(K mol), then the solution is:

```
>> Eₐ=75e3;
>> A=1e14;
>> T=300;
>> R=8.314;
>> k=A*exp(-Eₐ/(R*T))
k =
 8.7271
>>
```

2.2 Vectors, matrices and arrays

In the above, the single variables were used usually in scalar form. In MATLAB® this means that the variable is a 1 × 1 matrix. Two-dimensional matrices and arrays represent a numerical table, but mathematical operations with a matrix are applied in accordance with the rules of linear algebra, while arrays are used in element-wise operations.

2.2.1 Generation of vectors and matrices: vector and matrix operators

2.2.1.1 Generation of vectors

Vectors are presented as numbers written sequentially in a row or in a column, and termed, respectively, row or column vectors. They can also be presented as lists of words or equations. In MATLAB® a vector is generated by typing the numbers in square brackets with spaces or comma between them in the case of a row vector, and with semicolons between them, or by pressing Enter between them, in the case of a column vector.

The data for an aerobic biomass process as per Table 2.2 can be presented as two vectors, for example:

Table 2.2 Biomass data

Time (min)	0	25	50	75	100	125	150	200
Biomass (g/l)	5.15	5.21	5.52	6.55	7.15	7.75	7.59	7.45

```
>> time=[0 25 50 75 100 125 150 200]
time =
    0   25   50   75   100   125   150   200
>> b_mass=[5.15;5.21;5.5;6.55;7.15;7.75;7.59;7.45]
b_mass =
   5.1500
   5.2100
   5.5000
   6.5500
   7.1500
   7.7500
   7.5900
   7.4500
```

The data from the first line of Table 2.2 is assigned to the row vector named time

The data from the second line of Table 2.2 is assigned to the column vector named b_mass

There are also two frequently used operators for generating vectors, namely : (colon) and linspace.

The colon operator has the form

$$vector\_name=i:j:k$$

where i and k are respectively the first and last term in the vector and j is the step between the terms within it. The last number cannot exceed the last number k. The step for j can be omitted; in such a case it is equal to 1 by default. Examples are:

```
>> time=0:25:200
time =
    0   25   50   75   100  125  150  175  200
```
First number 0, last number 200, step 25

```
>> x=-3:7
x =
   -3  -2  -1   0   1   2   3   4   5   6   7
```
First number -3, last number 7, step by default is 1

```
>> y=0.2:0.12:1
y =
    0.2000   0.3200   0.4400   0.5600   0.6800   0.8000   0.9200
```
First number 0.2, last number 1, step 0.12

```
>> z=15.2:-3.21:1.3
z =
   15.2000  11.9900   8.7800   5.5700   2.3600
```
First number 15.2, last number 1.3, step -3. 21

The linspace operator has the form

$$vector\_name = linspace(a, b, n)$$

where a is the first number, b is the last number and n is the amount of numbers. When n is not specified, it takes the value of 100 by default.
For example:

```
>> x=linspace(0,28,8)
x =
    0    4    8   12   16   20   24   28
```
8 numbers, first number 0, last number 28

```
>> y=linspace(-10,100,3)
y =
   -10   45   100
```
3 numbers, first number -10, last number 100

```
>> z=linspace(15.2,1.3,5)
z =
   15.2000  11.7250   8.2500   4.7750   1.3000
```
5 numbers, first number 15.2, last number 1.3

```
>> v=linspace(0,100)
v =
  Columns 1 through 8
    0   1.0101   2.0202   3.0303   4.0404   5.0505   6.0606   7.0707
  ...
>>
```
Amount of numbers is omitted, default is 100, first number 0. last number 100

Published by Woodhead Publishing Limited

The position of an element in a vector is its address; for example, the fifth position in the eight-element vector b_mass above can be addressed as b_mass(5), and the element located here is 7.15. The last position in the b_mass vector may be addressed with the end terminator; for example, b_mass(end) is the last position in the b_mass vector and assigns the number located here, 7.45; another way to address the last element is to give the position number, namely b_mass(8).

2.2.1.2 Generation of matrices and arrays

A two-dimensional matrix or an array has rows and columns of numbers and resembles a numerical table, the only difference being in realization of certain mathematical operations. When the number of rows and columns is equal the matrix is square, rectangular otherwise. Like a vector, it is generated by typing the row of elements in square brackets with spaces or commas between them and with semicolons between the rows, or by pressing Enter between the rows; the number of elements in each row should be equal.

The elements can also be variable names or mathematical expressions.

As an example, Table 2.3 presents repeated tests on three batches of enzyme activity:

Table 2.3 Enzyme activity (mg⁻¹)

Batch 1	Batch 2	Batch 3
100.9	100.8	110.0
102.0	101.0	108.0
104.0	100.1	107.0

Matrix presentations of this table and other examples are:

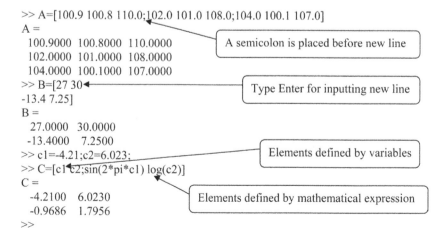

```
>> A=[100.9 100.8 110.0;102.0 101.0 108.0;104.0 100.1 107.0]
A =
  100.9000  100.8000  110.0000
  102.0000  101.0000  108.0000
  104.0000  100.1000  107.0000
>> B=[27 30
-13.4 7.25]
B =
  27.0000   30.0000
  -13.4000    7.2500
>> c1=-4.21;c2=6.023;
>> C=[c1 c2;sin(2*pi*c1) log(c2)]
C =
  -4.2100    6.0230
  -0.9686    1.7956
>>
```

A semicolon is placed before new line

Type Enter for inputting new line

Elements defined by variables

Elements defined by mathematical expression

For manipulations with matrix elements, row–column addressing is used. For instance, in matrix A in the previous example set A(2,3) refers to the number 108.0000 and A(3,2) to the number 100.1000. Row or column numbering begins with 1, such that the first element in matrix A is A(1,1).

For sequential elements or an entire row or column, the semicolon can be used; for example, A(2:3,2) refers to the second and the third numbers in column 2 of matrix A, A(:,n) refers to the elements of all rows in column n and A(m,:) to those of all the columns in row m.

In addition to row–column addressing, linear addressing can be used. In this case a single number is used instead of the row and column numbers, and the element's place within the matrix is indicated sequentially beginning from the first element of the first column and along it, then continuing along the second column and so up to the last element in the last column. For example, A(6) refers to element A(3,2), A(8) to A(2,3), A(4:6) is the same as A(:,2), etc.

Using square brackets, it is possible to generate a new matrix by combining an existing matrix with a vector or with another matrix.

```
>> V=linspace(127.1,252.3,3)                    Produce vector V by the linspace
V =
   127.1000  189.7000  252.3000
>> B=[A;V]                                      Create matrix B by joining matrix A
B =                                             and vector V, V added to A as new
   100.9000  100.8000  110.0000                 row
   102.0000  101.0000  108.0000
   104.0000  100.1000  107.0000
   127.1000  189.7000  252.3000
>> B(3,2)                                       Refer to the elements in row 3 and in
ans =                                           column 2 of the matrix B
   100.1000
>> B(2:4,1)                                     Refer to the elements in column 1 and
ans =                                           rows 2 through 4 of the matrix B
   102.0000
   104.0000
   127.1000

>> B(2,1:3)                                     Refer to the elements in row 2 and
ans =                                           columns 1 through 3 of the matrix B
   102   101   108
>> B(3,:)                                       Refer to the elements in row 3 and in
ans =                                           column 1 through 3 in matrix B
   104.0000  100.1000  107.0000
>> B(2:4)=6.4321                                Assign the value 6.4321 to the
B =                                             elements in column 1 and rows 2
   100.9000  100.8000  110.0000                 through 4
     6.4321  101.0000  108.0000
     6.4321  100.1000  107.0000
     6.4321  189.7000  252.3000
>>
```

Examples of this kind of matrix manipulation are presented below, using matrix A from the previous example.

For conversion of a row/column vector into a column/row vector and for the rows/columns exchange in matrices, the transpose operator ' (quote) is applied, for example:

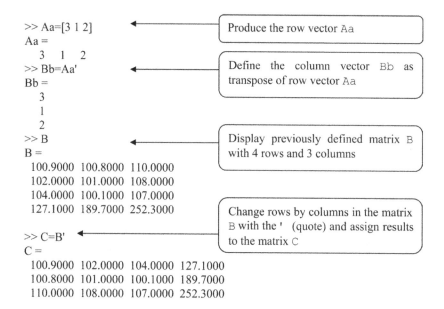

```
>> Aa=[3 1 2]
Aa =
   3   1   2
>> Bb=Aa'
Bb =
   3
   1
   2
>> B
B =
   100.9000 100.8000 110.0000
   102.0000 101.0000 108.0000
   104.0000 100.1000 107.0000
   127.1000 189.7000 252.3000

>> C=B'
C =
   100.9000 102.0000 104.0000 127.1000
   100.8000 101.0000 100.1000 189.7000
   110.0000 108.0000 107.0000 252.3000
```

Produce the row vector Aa

Define the column vector Bb as transpose of row vector Aa

Display previously defined matrix B with 4 rows and 3 columns

Change rows by columns in the matrix B with the ' (quote) and assign results to the matrix C

2.2.2 Matrix operations

Vectors, matrices and arrays can be used in various mathematical operations in the same way as single variables, as illustrated below.

2.2.2.1 Addition and subtraction

Addition and subtraction of two matrices are performed element by element, provided the matrices are equal in size; for example, when A and B are two matrices with size 3×2 each:

$$A = \begin{bmatrix} A_{11} & A_{12} \\ A_{21} & A_{22} \\ A_{31} & A_{32} \end{bmatrix} \quad \text{and} \quad B = \begin{bmatrix} B_{11} & B_{12} \\ B_{21} & B_{22} \\ B_{31} & B_{32} \end{bmatrix}$$

the sum of these matrices is

$$\begin{bmatrix} A_{11} + B_{11} & A_{12} + B_{12} \\ A_{21} + B_{21} & A_{22} + B_{22} \\ A_{31} + B_{31} & A_{32} + B_{32} \end{bmatrix}$$

In addition and subtraction operations the commutative law is valid, namely $A + B = B + A$.

2.2.2.2 Multiplication

Multiplication of matrices is more complicated; in accordance with the rules of linear algebra, it is feasible only when the number of row elements in the first matrix equals to that of column elements in the second – in other words, the inner matrix dimensions must be equal. Thus the above matrices A, 3 × 2, and B, 3 × 2, cannot be multiplied, but if B is replaced by another with size 2 × 3, the inner matrix dimensions are equal and multiplication becomes possible.

$$\begin{bmatrix} A_{11}B_{11} + A_{12}B_{21} & A_{11}B_{12} + A_{12}B_{22} & A_{11}B_{13} + A_{12}B_{23} \\ A_{21}B_{11} + A_{22}B_{21} & A_{21}B_{12} + A_{22}B_{22} & A_{21}B_{13} + A_{22}B_{23} \\ A_{31}B_{11} + A_{32}B_{21} & A_{31}B_{12} + A_{32}B_{22} & A_{31}B_{11} + A_{32}B_{23} \end{bmatrix}$$

It is not difficult to verify that the product $B*A$ is not the same as $A*B$, and the commutative law does not apply here.

Various examples of matrix addition, subtraction and multiplication are given below, using the same A and B matrices as in the preceding section.

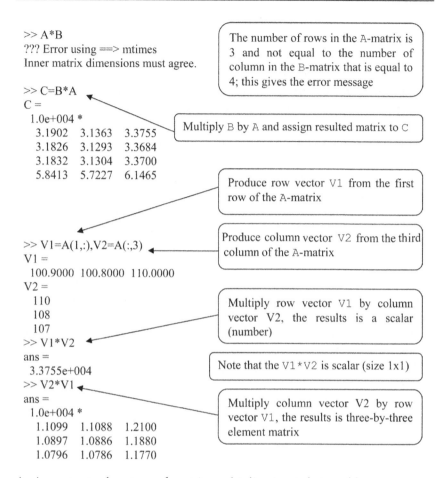

```
>> A*B
??? Error using ==> mtimes
Inner matrix dimensions must agree.

>> C=B*A
C =
  1.0e+004 *
    3.1902  3.1363  3.3755
    3.1826  3.1293  3.3684
    3.1832  3.1304  3.3700
    5.8413  5.7227  6.1465



>> V1=A(1,:),V2=A(:,3)
V1 =
  100.9000  100.8000  110.0000
V2 =
  110
  108
  107
>> V1*V2
ans =
  3.3755e+004
>> V2*V1
ans =
  1.0e+004 *
    1.1099  1.1088  1.2100
    1.0897  1.0886  1.1880
    1.0796  1.0786  1.1770
```

The number of rows in the A-matrix is 3 and not equal to the number of column in the B-matrix that is equal to 4; this gives the error message

Multiply B by A and assign resulted matrix to C

Produce row vector V1 from the first row of the A-matrix

Produce column vector V2 from the third column of the A-matrix

Multiply row vector V1 by column vector V2, the results is a scalar (number)

Note that the V1*V2 is scalar (size 1x1)

Multiply column vector V2 by row vector V1, the results is three-by-three element matrix

An important advantage of matrix multiplication is being able to present a set of linear equations in matrix form. For example, the set of two equations with two variables

$$A_{11}x_1 + A_{12}x_2 = B_1$$
$$A_{21}x_1 + A_{22}x_2 = B_2$$

may be written in compact matrix form as AX = B or in full matrix form as

$$\begin{bmatrix} A_{11} & A_{12} \\ A_{21} & A_{22} \end{bmatrix} \begin{bmatrix} x_1 \\ x_1 \end{bmatrix} = \begin{bmatrix} B_1 \\ B_2 \end{bmatrix}$$

2.2.2.3 Division

Division of matrices is even more complicated than their multiplication, as a result of the above-mentioned non-commutative properties of the matrix. A full explanation can be found in books on linear algebra. Here the related operators are described in the context of their usage in MATLAB®.

Identity and inverse matrices are often used in dividing operators. An identity matrix I is a square matrix whose diagonal elements are 1's and the remainder 0's. It can be generated with the eye command (see Table 2.4 on p. 26). The commutative law applies – multiplication of A by I or I by A yields the same result: AI = IA = A.

The matrix B is called the inverse of A when left or right multiplication leads to the identity matrix: AB = BA = I. The inverse matrix can be written as A^{-1}. In MATLAB® this can be written in two ways: B = A^-1 or with operator inv as B = inv(A).

Where matrix products are involved, left, \, or right, /, division is used. For example, to solve the matrix equation AX = B, with X and B column vectors, left division should be used: X = A\B, while to solve XC = B, with X and B row vectors and C as the transposed matrix of A, right division should be used: X = B/C.

For example, we can use matrix division to solve the following set of equations:

$$x_1 - 2x_2 = 8$$
$$6x_1 + 8x_2 = 12$$

According to the above, this set of equations can be represented in two matrix forms:

$$AX = B \text{ with } A = \begin{bmatrix} 1 & -2 \\ 6 & 8 \end{bmatrix}, B = \begin{bmatrix} 8 \\ 12 \end{bmatrix} \text{ and } X = \begin{bmatrix} x_1 \\ x_2 \end{bmatrix}, \text{ or}$$

$$XC = D \text{ with } C = \begin{bmatrix} 1 & 6 \\ -2 & 8 \end{bmatrix}, B = \begin{bmatrix} 8 & 12 \end{bmatrix} \text{ and } X = \begin{bmatrix} x_1 & x_2 \end{bmatrix}.$$

The solutions for both these forms are:

```
>> A=[1 -2;6 8];
>> B=[8;12];
>> X_left=A\B
X_left =
    4.4000
   -1.8000
```

Equation form AX=B
Solution with left division A\B

Equation form XC=D
Solution with right division D/C

```
>> C=A';
>> D=B';
>> X_right=D/C
X_right =
    4.4000   -1.8000
```

Note: in the latter case C is the transposed A matrix, D is the transposed B vector, and the solution X is a row vector

An application example with matrix division is given in Subsection 2.3.4.1.

2.2.3 Array operations

All previously described operations concern matrices obeying linear algebra rules; however, there are many calculations (in particular in bioscience) where the operations are carried out by the so-called element-by-element procedure. In these cases, to avoid confusion, we use the term 'array'. These element-wise operations are carried out with elements in identical positions in the arrays. In contrast to matrix operations, element-wise operations are confined to arrays of equal size; they are denoted with a point typed preceding the arithmetic operator, namely .* (element-wise multiplication); ./ (element-wise right division), .\ (element-wise left division) and .^ (element-wise exponentiation).

For example, if we have vectors $a = [a_1\ a_2\ a_3]$ and $b = [b_1\ b_2\ b_3]$ then element-by-element multiplication a.*b, a./b and exponentiation a.^b yields:

$$a.*b = [a_1\ b_1\ a_2\ b_2\ a_3\ b_3],\ a./b = [a_1/b_1\ a_2/b_2\ a_3/b_3],\ \text{and}\ a.^b = [a_1^{b_1}\ a_2^{b_2}\ a_3^{b_3}]$$

The same manipulations for two matrices $A = \begin{bmatrix} A_{11} & A_{12} & A_{13} \\ A_{21} & A_{22} & A_{23} \end{bmatrix}$ and

$B = \begin{bmatrix} B_{11} & B_{12} & B_{13} \\ B_{21} & B_{22} & B_{23} \end{bmatrix}$ lead to:

$$A.*B = \begin{bmatrix} A_{11}B_{11} & A_{12}B_{12} & A_{13}B_{13} \\ A_{21}B_{21} & A_{22}B_{22} & A_{23}B_{23} \end{bmatrix}, \qquad A./B = \begin{bmatrix} A_{11}/B_{11} & A_{12}/B_{12} & A_{13}/B_{13} \\ A_{21}/B_{21} & A_{22}/B_{22} & A_{23}/B_{23} \end{bmatrix}$$

and
$$A.\wedge B = \begin{bmatrix} A_{11}^{B_{11}} & A_{12}^{B_{12}} & A_{13}^{B_{13}} \\ A_{21}^{B_{21}} & A_{22}^{B_{22}} & A_{23}^{B_{23}} \end{bmatrix}$$

Element-wise operators are frequently used to calculate a function at series of values of its argument. Examples of array operations are:

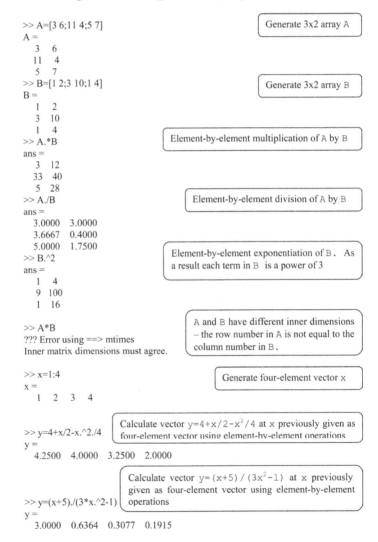

```
>> A=[3 6;11 4;5 7]
A =
    3    6
   11    4
    5    7
```
Generate 3x2 array A

```
>> B=[1 2;3 10;1 4]
B =
    1    2
    3   10
    1    4
```
Generate 3x2 array B

```
>> A.*B
ans =
    3   12
   33   40
    5   28
```
Element-by-element multiplication of A by B

```
>> A./B
ans =
    3.0000   3.0000
    3.6667   0.4000
    5.0000   1.7500
```
Element-by-element division of A by B

```
>> B.^2
ans =
    1    4
    9  100
    1   16
```
Element-by-element exponentiation of B. As a result each term in B is a power of 3

```
>> A*B
??? Error using ==> mtimes
Inner matrix dimensions must agree.
```
A and B have different inner dimensions – the row number in A is not equal to the column number in B.

```
>> x=1:4
x =
    1    2    3    4
```
Generate four-element vector x

```
>> y=4+x/2-x.^2./4
y =
    4.2500   4.0000   3.2500   2.0000
```
Calculate vector $y = 4+x/2-x^2/4$ at x previously given as four-element vector using element-by-element operations

```
>> y=(x+5)./(3*x.^2-1)
y =
    3.0000   0.6364   0.3077   0.1915
```
Calculate vector $y = (x+5)/(3x^2-1)$ at x previously given as four-element vector using element-by-element operations

2.2.4 Commands for generation of special matrices and some additional commands for matrices and arrays

There are also some commands for generating matrices with special values and those with random values. The `ones(m,n)` and `zeros(m,n)` commands are used for matrices of m rows and n columns with 1 and 0 as all elements. Various practical problems involve random numbers, for which the `rand(m,n)` or `randn(m,n)` command should be used, the former yielding a uniform distribution of elements between 0 and 1 and the latter a normal one with mean 0 and standard deviation 1. The `randseq` command, which generates random DNA, RNA and amino-acid sequences, is explained in Chapter 7. For generating a square matrix (n × n), these commands can be abbreviated to `rand(n)` and `randn(n)`. Examples are:

`>> ones(2,3)`
`ans =`
  1 1 1
  1 1 1
	Generate 2x3 matrix, in which all elements are equal to 1

`>> zeros(3,2)`
`ans =`
  0 0
  0 0
  0 0
	Generate 2x3 matrix, in which all elements are equal to 0

`>> a=rand(2,3)`
`a =`
  0.9501 0.6068 0.8913
  0.2311 0.4860 0.7621
	Generate 2x3 matrix a with uniformly distributed random numbers between 0 and 1

`>> v=rand(1,3)`
`v =`
  0.4565 0.0185 0.8214
	Generate row vector v with three uniformly distributed random numbers between 0 and 1

`>> b=randn(2,3)`
`b =`
  -0.4326 0.1253 -1.1465
  -1.6656 0.2877 1.1909
	Generate 2x3 matrix b with normally distributed random numbers

`>> w=randn(3,1)`
`w =`
  1.1892
  -0.0376
  0.3273
	Generate 3x1 vector w with three normally distributed random numbers

Integer random numbers can be generated with the `randi`-command as shown in Table 2.4.

In addition to the commands described in the previous sections, MATLAB® has many others that can be used for manipulation, generation and analysis of matrices and arrays; some of these are listed in Table 2.4.

Table 2.4 Commands for matrix manipulations, generation and analysis

Form of MATLAB® presentation	Description	MATLAB® example (inputs and outputs)
length(x)	Returns the length of vector x	>> x = [3 7 1]; >>length(x) ans 3|
size(a)	Returns two-element row vector; the first element is the number of rows in matrix a and the second the number of columns.	>>a = [1 2; 7 3; 9 6]; >>size(a) ans = 3 2
reshape(a,m,n)	Returns an m by n matrix whose elements are taken column-wise from a. Matrix a must have m*n elements	>>reshape(a,2,3) ans = 1 9 3 7 2 6
strvcat(t1,t2,t3,…)	Generates the matrix containing the text strings t1, t2, t3, … as rows.	>> t1 = 'Alanine'; >> t2 = 'Arginine'; >> t3 = 'Asparagine'; >> strvcat(t1,t2,t3) ans Alanine Arginine Asparagine
zeros(m,n)	Generates an m by n matrix of all zeros	>> zeros(2,3) ans = 0 0 0 0 0 0
diag(x)	Generates a matrix with elements of vector x placed along the diagonal	>> x = 1:3;diag(x) ans = 1 0 0 0 2 0 0 0 3
eye(n)	Generates a square matrix with diagonal elements 1 and others 0	>> eye(4) ans = 1 0 0 0 0 1 0 0 0 0 1 0 0 0 0 1

Published by Woodhead Publishing Limited

Table 2.4 *Continued*

randi(imax,m,n)	Returns an m by n matrix of integer random numbers from value 1 up to imax, the maximal integer value	>> randi(10,1,3) ans = 9 10 2
b = min(a)	Returns row vector b with minimal numbers of each column in the matrix a. If a is vector, b is equal to the minimal number in a	>> a = [1 2; 7 3; 9 6]; >> b = min(a) b = 1 2 >> a = [1 2 7 3 9 6]; >> b = min(a) b = 1
b = max(a)	Analogously to min but for maximal element	>> a = [1 2 7 3 9 6]; >> b = max(a) b = 9
b = mean(a)	Returns row vector b with mean values calculated for each column of the matrix a. If a is a vector, returns average value of the vector a	>> a = [1 2; 7 3; 9 6]; >> b = mean(a) b = 5.6667 3.6667
sum(a)	Returns row vector b with column sums of matrix a. Returns vector sum if a is a vector	>> a = [1 2; 7 3; 9 6]; >> sum(a) ans = 17 11
std(a)	Analogously to sum but calculates standard deviation	>> a = [1 2; 7 3; 9 6]; >> std(a) ans = 4.1633 2.0817
det(a)	Calculates determinant of the square matrix a	>> a = [5 6;12 1]; >> det(a) ans = -67
sort(a)	For vector or matrix. Sorts, respectively, elements of a vector or each column of a in ascending order.	>> a = [5 6;12 1;1 7]; >> sort(a) ans = 1 1 5 6 12 7
num2str(a)	Converts a single number or numerical matrix elements into a string representation	>> a = 12.4356; >> num2str(a) ans = 12.4356

All matrices described above had numerical elements even if they have the expressions as an element because these expressions yield numbers, when evaluated. But as a single element or number of elements of a matrix, a string(s), can be used. A string is an array of characters – letters and/or symbols. A string is entered in MATLAB® between single quotes, e.g. 'Protein' or 'Human cells DNA <deoxyribonucleic acid> totals about 3 meters in length'. Each character of the string is presented and stored as a number (thus the set of characters represents a vector or an array) and can be addressed as an element of a vector or array, e.g. a(5) in the string 'Protein' is the letter 'e'. Some examples with string manipulations are:

```
>> a='Protein'
a =
Protein
```

Assign the string 'Protein' to the variable a; it takes 7 letters and is a 7-element row vector

```
>> a(4)
ans =
t
```

The fourth element of the vector a is the letter t, thus a(4) is t

```
>> a(5:7)
ans =
ein
```

The fifth, sixth, and seventh elements of the vector a are the letters e, i, and n

```
>> a([1 3 2 4])
ans =
Port
```

The 1st, 3rd, 2nd, and 4th elements of the vector a are the letters P, o, r, and t

Strings can be placed as elements in a vector or a matrix. String rows are divided the same as numerical rows by a semicolon (;) and strings within the rows by a space or a comma. Rows should have the same number of elements and each column element must be the same length as the longest of the column elements. To achieve this alignment, spaces should be added to shorter strings; for example

> Error due to inequality in length of the strings:
> Number of the characters in the word DNA is
> 3 and in Adenine- and Thymine is 7

```
>> Name=['DNA';'Adenine';'Thymine']
??? Error using ==> vertcat
CAT arguments dimensions are not consistent.
```

> Four spaces were added after DNA; now the
> length of each of the strings is the same and
> the three strings are successfully written as
> column vector

```
>> Name=['DNA    ';'Adenine';'Thymine']
Name =
DNA
Adenine
Thymine
```

2.2.5   Application examples

2.2.5.1   DNA bases table

DNA has four bases, adenine, cytosine, guanine and thymine.

Problem: Generate and display the matrix in which the first column is the serial number and the second is the base name.
  Make it in the following steps:

- Generate a numerical column of values from 1 to 4.
- Generate a string column with the names adenine, cytosine, guanine and thymine.
- Join these two columns in to a matrix. In MATLAB® all matrix elements should be of the same type; for example, if strings are written in one column, then another column must also contain strings or vice versa – in our case we use the num2str command, which transforms numerical data into string data.
- Display table title 'DNA bases'.
- Display the resulting matrix.

The commands are:

>> No=[1:4]';

> Generate numerical column vector No

>> Name=strvcat(' Adenine',' Cytosine',' Guanine',' Thymine');

> Generate string column vector Name

>> Table=[num2str(No) Name];

> Generate **Table** as two column string matrix. The `num2str` command transforms numbers to string

>> disp(' '),disp(' DNA bases'),disp(Table)

```
DNA bases
1  Adenine
2  Cytosine
3  Guanine
4  Thymine
```

> The first `disp` displays blank line; the second `disp` displays the title; the third `disp` displays the table

2.2.5.2 Blossom statistics

A biotech company has developed a new type of fruit tree. Below are blossom yield data from ten trees, checked twice at different times, and presented in two rows (time) and in ten columns (trees):

$$27 \;\; 27 \;\; 35 \;\; 28 \;\; 32 \;\; 33 \;\; 31 \;\; 35 \;\; 28 \;\; 30$$

$$32 \;\; 35 \;\; 34 \;\; 33 \;\; 36 \;\; 35 \;\; 31 \;\; 27 \;\; 28 \;\; 35$$

Problem: Find the mean, the difference between maximal and minimal values (range), and the standard deviation for every row of the data, and display the results to two decimal places, using the `fprintf` command.

The steps are as follows:

- The tree blossom data are assigned to a two-row matrix.
- The mean, range and standard deviation are calculated by the appropriate MATLAB® commands.
- The statistics obtained are displayed via the `fprintf` command.

The commands are:

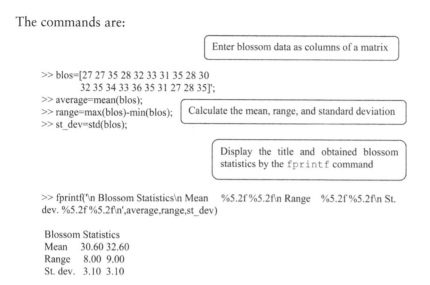

```
>> blos=[27 27 35 28 32 33 31 35 28 30
          32 35 34 33 36 35 31 27 28 35]';
>> average=mean(blos);
>> range=max(blos)-min(blos);
>> st_dev=std(blos);
```

```
>> fprintf('\n Blossom Statistics\n Mean    %5.2f %5.2f\n Range    %5.2f %5.2f\n St.
dev. %5.2f %5.2f\n',average,range,st_dev)
```

```
Blossom Statistics
Mean    30.60 32.60
Range   8.00  9.00
St. dev. 3.10  3.10
```

2.2.5.3 Wind chill index

Wind chill w is the apparent temperature experienced at wind velocity v and air temperature T. An empirical formula for it is:

$$w = 91.4 + 0.0817\left(3.71\sqrt{v} + 5.81 - 0.25 * v\right)(T - 91.4)$$

where w is in degrees Fahrenheit, v is in m/s (4, 5, 10, 15, 20, 25, 30, 35, 40 and 45) and T decreases from 35 to −35 at steps of 10°, also in Fahrenheit.

Problem: Write the command for the wind chill matrix by giving vectors v and T and display it as a table, so that w is arranged along the rows at constant v and along the columns at constant T.

The steps are as follows:

• Generate separately the column-vector v (size 10 × 1) and the row-vector T (size 1 × 8).
• Calculate the w-matrix according to the formula. The first multiplier (the terms in the first set of parentheses of the w-expression) has size 10 × 1 and the second (the terms in the second set of parentheses) has size 1 × 8, and thus their product according to linear algebra would be size 10 × 8 (extreme values of the sizes of these vectors).
• Display the title 'Wind chill' and w-matrix with the digits that are before the decimal point only.

The commands to the solution are:

> Enter the velocity values in a column vector and the temperature values in arrow vector

```
>> v=[4,5:5:45]';T=35:-10:-35;
>> w=round(91.4+0.0817*(3.71*sqrt(v)+5.81-.25*v)*(T-91.4));
```

> Calculate the wind chill index

> Display title for resulting table using disp

```
>> disp(' '),disp('            Wind Chill'),...
```

> Display calculated wind chill indices using fprintf

```
fprintf('%5.0f%5.0f%5.0f%5.0f%5.0f%5.0f%5.0f%5.0f\n',w')
   35    25    15     5    -5   -15   -25   -35
   32    22    11     1   -10   -20   -31   -41
   22    10    -2   -15   -27   -39   -52   -64
   16     2   -11   -25   -38   -51   -65   -78
   11    -3   -17   -31   -46   -60   -74   -88
    8    -7   -22   -36   -51   -66   -81   -96
    6   -10   -25   -40   -55   -71   -86  -101
    4   -12   -27   -43   -58   -74   -89  -105
    3   -13   -29   -45   -60   -76   -92  -108
    2   -14   -30   -46   -62   -78   -94  -109
```

2.2.5.4 Weight versus height

Measurements of students' weight w (in kg) and height h (in cm) in a group at an American college showed the following results: heights – 155, 175, 173, 175, 173, 162, 173, 188, 190, 173, 173, 185, 178, 168, 162, 185, 170, 180, 175, 180, 175, 175, 180, 165; weights – 54, 66, 66, 71, 68, 53, 61, 86, 92, 57, 59, 80, 70, 59, 50, 145, 68, 78, 67, 90, 75, 74, 84, 53.

These data were fitted by the polynomial expression

$$w_f = 906.14 - 11.39h + 0.03780h^2.$$

Problem: Write the commands to input data in a two-row matrix, w_h, and calculate the weight by the expression and the percentage error, $100(w - w_f)/w$. Display the results as a three-column table listing every third value of w, w_f and the error.

The steps are as follows:

- The height and weight values are assigned as above.
- The weights w_f and the errors are calculated according to the described expressions.

Published by Woodhead Publishing Limited

- Every third value of the input and calculated values of weight, and the calculated errors, are written in a three-row matrix *tab* and displayed as a three-column table.

> Enter the values of height in the first row, and the values of weight in the second row of the matrix w_h.
> Note, three periods (...) are entered for continuing the data in the next line

```
>> w_h=[155,175,173,175,173,162,173,188,190,173,173,185,178,168,162,...
185,170,180,175,180,175,175,180,165;
54,66,66,71,68,53,61,86,92,57,59,80,70,59,50,145,68,78,67,90,75,74,84,53];
```

> Calculate the weight by the expression w(h)

```
>> w_f=906.14-11.39*w_h(1,:)+0.03780*w_h(1,:).^2;
```

```
>> error=100*(w_h(2,:)-w_f)./w_h(2,:);
```

> Calculate the error

```
>> pr=1:3:length(w_f);
```

> Prepare three-row matrix with every third element of the imputed and calculated weights, and the error

```
>> tab=[w_h(2,pr);w_f(pr);error(pr)];
```

```
>> fprintf('%8.0f %8.0f %8.0f\n',tab)
      54      49      10
      71      71       1
      61      67     -10
      57      67     -18
      70      76      -9
     145      93      36
      67      71      -5
      74      71       5
```

> Print the resulting table
> Note, the rows of the tab are printed as columns by the fprintf command

2.3 Flow control

A calculation program represents a sequence of commands implemented in a given order. However, there are many cases when the written order of single – or group of – commands should be altered, for example when a calculation should be repeated with new parameters, or when one of several expressions has to be chosen to calculate a variable. For example, bacterial growth proceeds in four phases, lag, log, stationary and death, for each of which a different expression should be used to calculate population size. Another example is provided by DNA, RNA and protein sequence analyses, where the alignment procedure is repeated until the best matching score is reached.

Flow control is applied in such processes. In MATLAB®, special commands, usually called conditional statements, are used for these purposes; by this method the computer decides which command should be carried out next. The most frequent flow control commands are described below.

2.3.1   Relational and logical operators

Relational operators. Operators matching a pair of values are called relational or comparison operators; the application result of such an operator is written as 1 (true value) or 0 (false value) – for example, the expression $x < 3$ results in 1 if x is less than 3 and in 0 otherwise.

The relational operators are: < (less than), > (more than), <= (less than or equal to), >= (more than or equal to), = = (equal to) and ~= (not equal to). Two-sign operators should be written without spaces.

Where a relational operator is applied to a matrix or an array, it performs element-by-element comparisons. This returns an array of ones where the relation is true (the array has the same size as the size of the matrices compared), and zeros where it is not. If one of the compared objects is scalar and the other a matrix, the scalar is matched against every element of the matrix. The ones and zeroes are logical data and are not the same as numerical data, although they can be used in arithmetical operations.

Some examples are:

```
>> 2*2==12/3
ans =
   1
```
Since of 2x2 is identical with 12/3 the result is true, thus the answer is 1

```
>> sin(2*pi)~=1
ans =
   1
```
Since sin 2 =0 and not 1 the result is true, thus the answer is 1

```
>> M=[-7 8 -15;7 -8 4;-2 -15 -2]
M =
   -7    8   -15
    7   -8    4
   -2  -15   -2
```
Produces a 3x3 matrix M

```
>> B=M<=0
B =
   1   0   1
   0   1   0
   1   1   1
```
Checks whether each element in the matrix M is less than 0

```
>> M(B)
ans =
   -7
   -2
   -8
  -15
  -15
   -2
```
Display the elements of M which are less than 0

Note: displayed result is a vector that contains the elements of M in positions where B is true (logical 1)

Published by Woodhead Publishing Limited

Logical operators. Logical operators are designed for operations with the true or false values within the logical expressions. They can be used as addresses in another vector, matrix or array.

In MATLAB® there are three logical operators: & (logical AND), | (logical OR) and ~ (logical NOT). Like the relational operators they can be used as arithmetical operators and with scalars, matrices and arrays. Comparison is element-by-element with logical 1 or 0 when the result is true or false, respectively. MATLAB® also has equivalent logical functions: and(A,B), equivalent to A&B, or(A,B), equivalent to A|B, and not(A,B), equivalent to A~B. If the logical operators are performed on logical variables the results will follow Boolean algebra rules. In operations with logical and/or numerical variables the results are logical 1 or 0.

Some examples are:

```
>> x=-1.5;
```
Define variable x

```
>> -2<x<-1
ans =
    0
```
The statement leads to an incorrect answer as it runs from left to right. −2<x is true (1), then 1<−1 is false (0)

```
>> x>-2&x<-1
ans =
    1
```
Here the logical & is used and leads to a correct result. First the inequalities are run, both are true (1), then the & which leads to the answer 1

```
>> ~(x<3)
ans =
    0
```
x<3 is stated in the parentheses and is run first, is true (1), then ~1 is 0

```
>> ~x<3
ans =
    1
```
Here ~x is executed first, x is nonzero then true (1), ~1 is 0, and 0<3 is true

Another MATLAB® logical function is find, which in it simplest forms reads as

$$i = find(x) \quad or \quad i = find(A>c)$$

where *i* is a vector of elements which are non-zero (first case), or belong to a vector A and are larger than *c* (second case); for example, vector *v* = [12 0.1 3.4 0 -2.5], and thus

```
>> i=find(v)
i =
    1   2   3   5
>> find(v>1)
ans =
    1  3
```

Note: any relational operator(s) can be written in the find command, e.g. find(A<0.5), or find(A>=9). The order in which combinations of relational, logical and conditional operators is executed (so-called precedence rules) is obtainable in advanced MATLAB® courses. The order of execution of such an operator can also be annotated using parentheses.

2.3.1.1 Application example: molecular weight screening

The molecular weights of 20 randomly generated sequences comprising 20 letters of the DNA alphabet are 1558, 1758, 1794, 1480, 1712, 1738, 1636, 1546, 1688, 1648, 1654, 1676, 1616, 1726, 1760, 1592, 1634, 1742, 1652 and 1740.

Problem: Use relational and logical operators to determine the number of sequences with molecular weight (a) less than 1650, (b) between 1650 and 1700, and (c) more than 1700. Display the molecular weights for each of these groups. The commands that solved this problem are presented on p. 37.

2.3.2 If statements

In program flow control, various conditional statements are widely used. The first of them is the if statement, which has three forms: if ... end, if ... else ... end, and if ... elseif ... else ... end. Each if construction terminates with the word end; its words appear on the screen in blue.

The if statements are shown in Table 2.5.

In conditional expressions the relational and logical operators are used, for example a< = b&a> = c or b = = c.

When the if with a conditional expression is typed next to the prompt >> and Enter is pressed, the next (and each additional) line appears without the prompt until the word end is typed.

An application example in which the if statement is used is presented at the end of Section 2.3.

2.3.3 Loops in MATLAB®

Another method of program flow control is a loop, which permits a single command, or a group of commands, to be repeated several times. Each

Define vector m_weight with the molecular weight data

```
>> m_weight=[1558 1758 1794 1480 1712 1738 1636 1546 1688 ...

1648 1654 1676 1616 1726 1760 1592 1634 1742 1652 1740];

>> m_less1650=find(m_weight<1650);
```

Find addresses (indices) where the molecular weight is less than 1650

```
>> N_m_less1650=sum(m_weight<1650)
N_m_less1650 =

    8
```

Calculate amount of sequences with weights less than 1650

```
>> m_weight(m_less1650)

ans =
```

Display the weights less than 1650

```
    1558    1480    1636    1546    1648    1616    1592    1634
```

Find addresses with molecular weights between 1650 and 1700

```
>> m_between1650and1700=find(m_weight>=1650&m_weight<=1700)

m_between1650and1700 =
```

Calculate amount of sequences with weights between 1650 and 1700

```
    9   11   12   19
>> N_m_between1650and1700=sum(m_weight>=1650&m_weight<=1700)
N_m_between1650and1700 =

    4
```

```
>> m_weight(m_between1650and1700)

ans =
```

Display the weights between 1650 and 1700

```
    1688    1654    1676    1652
```

Find addresses with molecular weights larger than 1700

```
>> m_above1700=find(m_weight>1700);
>> N_above1700=sum(m_weight>1700)
```

Calculate amount of sequences with weights larger than 1700

```
N_above1700 =

    8
```

```
>> m_weight(m_above1700)
```

Display the weights larger than 1700

```
ans =

    1758    1794    1712    1738    1726    1760    1742    1740
```

Table 2.5 If statements

if ... end	if ... else ... end	if ... elseif ... else ... end
if conditional expression MATLAB® command(s) end	if conditional expression MATLAB® command(s) else MATLAB® command/s end	if conditional expression MATLAB® command(s) elseif conditional expression MATLAB® command(s) else MATLAB® command(s) end

cycle of commands is termed a pass. There are two loop commands in MATLAB®: `for ... end` and `while ... end`. These words appear on the screen in blue. As with `if` statements, each `for` or `while` construction should terminate with the word `end`.

Table 2.6 Loops

for ... end loop	while ... end loop
for k = [initial : step : final]	While conditional expression
MATLAB® command(s)	MATLAB® command(s)
end	end.

The loop statements are written in general form in Table 2.6.

In `for ... end` loops the commands written between `for` and `end` are repeated k times, a number which increases for every pass by addition of the `step`-value; this process is continued until k reaches or exceeds the final value.

The square brackets in the expression for k (Table 2.6) mean that k can be assigned as a vector, for example k = [3.5 –1.06 1:2:6]. The brackets can be omitted if there are only colons in k, e.g. k = 1:3:10. The last pass is followed by the command next to the loop. For some calculations realized with `for ... end` loops matrix operations can serve as well. In such cases the latter are actually superior, as the `for ... end` loops work slowly. The advantage is negligible for short loops with a small number of commands, but is appreciable for large loops with numerous commands.

The `while ... end` loop is used where the number of passes is not known in advance and the loop terminates only when the conditional expression is false. In each pass MATLAB® executes the commands written between the `while` and `end`; the passes are repeated until the conditional expression is true. An incorrectly written loop may continue indefinitely; for example,

```
> a = 3;
>> while a >2
a = 2*a
end
```

In this case the expression a = Inf appears repeatedly on the screen. To interrupt the loop, the `Ctrl` and `C` keys should be pressed together.

Examples of the `for ... end` and `while ... end` loops used for calculating e^x via the series $\sum_{k=0} \dfrac{x^k}{k!}$ at x = 2.3 are (see p. 39):

Published by Woodhead Publishing Limited

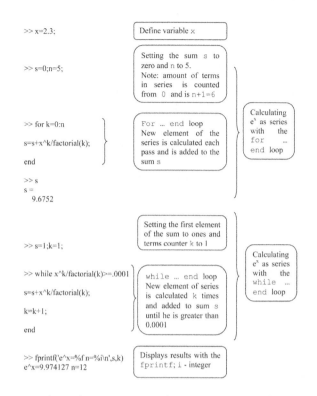

```
>> x=2.3;
```
Define variable x

```
>> s=0;n=5;
```
Setting the sum s to zero and n to 5. Note: amount of terms in series is counted from 0 and is n+1=6

```
>> for k=0:n

s=s+x^k/factorial(k);

end
```
For ... end loop New element of the series is calculated each pass and is added to the sum s

Calculating e^x as series with the `for` ... `end` loop

```
>> s
s =
   9.6752
```

```
>> s=1;k=1;
```
Setting the first element of the sum to ones and terms counter k to 1

```
>> while x^k/factorial(k)>=.0001

s=s+x^k/factorial(k);

k=k+1;

end
```
while ... end loop New element of series is calculated k times and added to sum s until he is greater than 0.0001

Calculating e^x as series with the `while` ... `end` loop

```
>> fprintf('e^x=%f n=%i\n',s,k)
e^x=9.974127 n=12
```
Displays results with the `fprintf`; i - integer

In the `for ... end` loop the sum s is calculated; at the start of the first pass $s = 0$, and during this pass the first term ($k = 0$) is calculated and added to s. On the second pass $k = k + 1 = 1$, the second term of the series is calculated and added to s. This procedure is repeated up to $k = n$ ($n = 5$ in this example). After this the loop ends and the value obtained is displayed after entering the variable name s. The number of passes is fixed in `for ... end` loops. In situations where the `while ... end` loop is used, a condition for loop ending must be given. A value of the kth term larger than 0.0001 was taken for this. With the start of the first pass the value of s is equal to a value of the first term of the series ($k = 0$, $x^0/0! = 1$) and k is equal to 1 (k in this case is called counter), and these values are assigned before the loop. In the first pass the second term of the series is calculated and added to the sum s, and the value of k is increased by 1. If the condition is true the next pass is started and the next term calculated; if it is false, the loop is ended and the `fprintf` command displays the value of s obtained and number of terms used – if the latter value is an integer then a conversion character i is used to display the value of k.

For ... end and `while ... end` loops and `if` statements can be included in other loops, in `if` statements, or one into another. There is no limit to the number of such inclusions.

Published by Woodhead Publishing Limited

2.3.4 Application examples

2.3.4.1 Compound concentration and the instrument response

An instrument used in a biotechnology laboratory has a response R with values 0.31, 0.43, 0.70, 1.1, 1.5, 1.79 and 2.2 to the following compound Z concentrations, c: 100, 150, 250, 400, 550, 650 and 850 mg/mL. The best linear fit equation $c = a_1 + a_2R$ is obtainable by solving the set

$$a_1 n + a_2 \sum_{i=1}^{n} R_i = \sum_{i=1}^{n} c_i$$

$$a_1 \sum_{i=1}^{n} R_i + a_2 \sum_{i=1}^{n} R_i^2 = \sum_{i=1}^{n} R_i c_i$$

where n is the number of (R,c) points.

This set can be represented in matrix forms AX = B or XA = B, in our case:

$$\begin{bmatrix} n & \sum_{i=1}^{n} R_i \\ \sum_{i=1}^{n} R_i & \sum_{i=1}^{n} R_i^2 \end{bmatrix} \begin{bmatrix} a_1 \\ a_2 \end{bmatrix} = \begin{bmatrix} \sum_{i=1}^{n} c_i \\ \sum_{i=1}^{n} R_i c_i \end{bmatrix} \quad \text{or} \quad \begin{bmatrix} a_1 & a_2 \end{bmatrix} \begin{bmatrix} n & \sum_{i=1}^{n} R_i \\ \sum_{i=1}^{n} R_i & \sum_{i=1}^{n} R_i^2 \end{bmatrix} = \begin{bmatrix} \sum_{i=1}^{n} c_i & \sum_{i=1}^{n} R_i c_i \end{bmatrix}$$

Problem: Define the a-coefficients with left- and right divisions; the result can be printed as a linear equation with the relevant coefficients a_1 and a_2.

The steps to be taken are as follows:

- Generate two row vectors with the given R and c values;
- Generate a 2 × 2 matrix A with the sums on the left-hand side of the first matrix forms; for sums the `sum` command can be used;
- Generate a column vector B with the sums on the right-hand side of the first matrix equation;
- Use the left division A\B for calculation of the a-coefficients;
- Display with the `fprintf` command the coefficients in the written linear equation;
- Resort to the right division A\B. For this A should be written as per the second matrix equation, using the quote operator ('). The letter serves to transform the column vector B into the row vector; the right division in this case is simply verification of the previous solution.
- Display with the `fprintf` command the coefficients directly in the written linear equation.

Published by Woodhead Publishing Limited

The commands for solution are:

>> R=[0.31,0.43,0.70,1.1,1.5,1.79,2.2];
>> c=[100 150 250 400 550 650 850];

> Generate vectors R and c of the response and concentration data

>> A=[length(R) sum(R);sum(R) sum(R.^2)];
>> B=[sum(c);sum(R.*c)];

> Generate the matrix A and vector B for solving the first matrix form: AX=B

>> a=A\B
a =
 -22.2676
 386.7837

> Solution via the left division: X=A\B

> Using fprintf for displaying results with 2 digits after decimal point

>> fprintf('\n The equation is c=%5.2f+%5.2f*R\n',a(1),a(2))

The equation is c=-22.27+386.78*R

>> aa=B'/A'

aa =

 -22.2676 386.7837

> Solution via the right division: X=B/A

> Note: for solving the second matrix form: XA=B, the matrix A and the vector B should be transposed by the quote operator (')

>> fprintf('\n The equation is c=%5.2f+%5.2f*R\n',aa(1),aa(2))

The equation is c=-22.27+386.78*R

2.3.4.2 Bacteria population growth

Changes in a bacteria population of size N (cells/mL) were measured from $t_0 = 1.5$ up to 14 (hours). The results obtained were fitted by the following expressions:

$$N = \begin{cases} \text{floor}\left[N_0 e^{\mu(t-1.5)}\right], & 0 \leq t < 8.5 \\ N_c, & 8.5 \leq t \leq 12 \\ \text{floor}\left[N_c e^{-2.338(t-12)}\right], & 12 < t \leq 14 \end{cases}$$

where t is time in hours, N_0 is the initial bacteria population at time t_0, N_c is the population in the stationary phase, μ is the growth rate constant (h^{-1}), floor means that bacteria number should be an integer; the parameters required are: $N_0 = 84$, $N_c = 35.6 \times 10^8$ and $\mu = 2.18$.

<u>Problem</u>: Calculate bacteria populations at 1.5, 2, 3, ... , 14 hours.
  The required steps are:

- Determine the variables N_0, N_c and μ, and a 1 × 14 vector with the t values;
- Use the `for... end` loop in which every pass runs for the new t value defined by its index (address) – $t(i)$;
- Introduce in the loop the statement `if ... elseif ... else ... end` in which the conditions and expressions on the right-hand side of the equation defining N should be written in the blank spaces; the values of N should be indexed to generate the vector of calculated N-values for each t;
- Display with the `fprintf` the vector of calculated N-values.

The commands for calculating N are

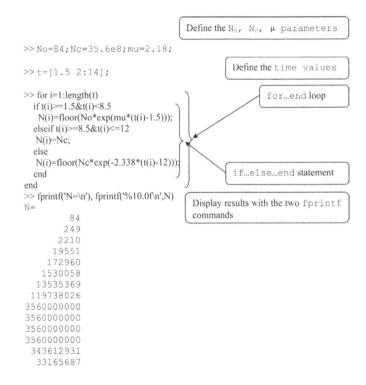

2.3.4.3 Dilution

The resulting molar concentration of a solution, M_2, is calculated via

$$M_2 = \frac{M_1 V_1}{V_2}$$

where V_2 is final solution volume, and M_1 and V_1 are the initial concentration and solution volume, respectively.

A series of standard solutions were prepared with M_1 values of 0.5, 1, 1.5 and 2 mol/L and common V_1 values of 0.1 L.

Problem: Calculate the table of molar concentrations M_2 if each of the prepared series of standard solutions was diluted by 0.1, 0.3, 0.6 and 0.9 L of water.

This can be constructed in two ways – using for... end loops and without the loops using the vectors for V_2 and M_1 only.

The required steps are as follows:

- Enter the value of V_1 and generate vectors for M_1 and V_2;
- Generate a matrix with the number of rows equal to the length of the M_1 vector and that of columns equal to the length of the V_2 vector; this step pre-allocates the matrix and serves to reduce operation time when the for ... end construction is used (a minor consideration for small matrices but quite significant for large ones);
- Calculate M_2 (with the expression above) in the two for ... end loops: the external for M_1 and the internal for V_2; such a construction yields all values M_2 for each V_2;
- Display the calculated matrix M_2 with the fprintf, in which the values obtained are presented with three digits after the decimal point.
- Repeat the same calculations but without loops; for this rewrite the expression in the form $M_1 V_1 (1/V_2)$ so that the element-wise division in brackets comes first and is followed by the multiplication; $(1/V_2)$ produces a row vector with size 1 × 4 and for the inner dimension equality the vector M_1 should be transformed by the quote operation (') to a column vector with size 4 × 1; the next multiplication by the scalar V_1 does changes the vector size, and the product of the [4 × 1]*[1 × 5] matrices is the final 4 × 5 matrix with the calculated M_2 values.

Published by Woodhead Publishing Limited

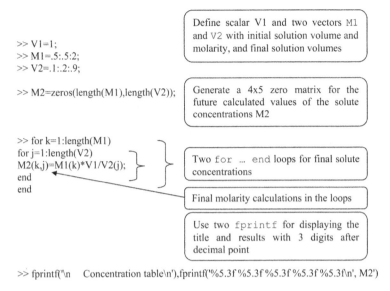

```
>> V1=1;
>> M1=.5:.5:2;
>> V2=.1:.2:.9;
```

Define scalar V1 and two vectors M1 and V2 with initial solution volume and molarity, and final solution volumes

```
>> M2=zeros(length(M1),length(V2));
```

Generate a 4x5 zero matrix for the future calculated values of the solute concentrations M2

```
>> for k=1:length(M1)
for j=1:length(V2)
M2(k,j)=M1(k)*V1/V2(j);
end
end
```

Two for ... end loops for final solute concentrations

Final molarity calculations in the loops

Use two fprintf for displaying the title and results with 3 digits after decimal point

```
>> fprintf('\n    Concentration table\n'),fprintf('%5.3f %5.3f %5.3f %5.3f %5.3f\n', M2')
```

```
Concentration table
5.000 1.667 1.000 0.714 0.556
10.000 3.333 2.000 1.429 1.111
15.000 5.000 3.000 2.143 1.667
20.000 6.667 4.000 2.857 2.222
```

Molarity calculations without loops

```
>> M2v=M1'*V1*(1./V2);
>> fprintf('\n    Concentration table\n'),fprintf('%5.3f %5.3f %5.3f %5.3f %5.3f\n',M2v')
```

```
Concentration table
5.000 1.667 1.000 0.714 0.556
10.000 3.333 2.000 1.429 1.111
15.000 5.000 3.000 2.143 1.667
20.000 6.667 4.000 2.857 2.222
```

Use two fprintf for displaying the title and results with 3 digits after decimal point

2.4 Questions for self-checking and exercises

1. Which command should be used to list the names of variables in the workspace? Choose from: (a) lookfor, (b) whos, (c) who.
2. The predefined variable π with format long is displayed as: (a) 3.1416, (b) 3.14, (c) 3.141592653589793, (d) 3.141592653589793e+000?
3. Which of the MATLAB® expressions below is suitable for the equation
$$y = \frac{\log 10}{\ln 2} ?$$
    (a) y = log(10)/ln(2), (b) y = log10(10)/ln(2),
    (c) y = log10(10)/log(2), (d) y = log(10)/log(2)
4. The command M = [1;2;3;4] generates (a) a row vector, (b) a column

vector or (c) a square matrix?

5. Which of the next special commands generates the matrix $\begin{bmatrix} 1 & 1 & 1 \\ 1 & 1 & 1 \\ 1 & 1 & 1 \end{bmatrix}$:

(a) `zeros(4,3)`, (b) `eye(3)`, (c) `ones(3)`, (d) `diag((1:4)./(1:4))`?

6. The command that generates uniformly distributed integer random numbers is: (a) `rand`, (b) `randn` or (c) `randi`?

7. Which of the answers below is correct for the division [2 4;6 8]/[1 2;3 4]:
    a) ans =   b) ans =
       2 2   2 0
       2 2   0 2

8. For transformation of a row vector T with numerical values into a column vector, the following MATLAB® expression should be written:
    (a) `1/T`, (b) `inv (T)`, (c) T⁻¹, (d) T′ ?

9. To address the element in a 3 × 3 matrix A located at the third row and the second column, the following MATLAB® expression should be written:
    (a) A(3,2),   (b) A(32),
    (c) A(6),   (d) A(2,3).
    Note: More than one correct answer is possible.

10. With some idealization, the corpuscle of muscle hemoglobin can be treated as a sphere 20 microns in diameter. Calculate its volume with the expression

$$V = \frac{4}{3}\pi\left(\frac{d}{2}\right)^3$$

11. The mass flow rate (L/min) of blood in the human circulatory system can be calculated by the empirical expression

$$\psi = 0.707 m^{0.425} h^{0.725} + 0.625$$

where m = mass in kg and h = height in m. Calculate the mass flow rate in men with parameters m = 80 kg and h = 1.8 m.

12. The temperature distribution in a biological system containing blood vessels is described by the expression

$$T = T_b + \left(T_a - T_b\right)\frac{\ln\dfrac{b}{r}}{\ln\dfrac{b}{a}}$$

Calculate the temperature at $r = (a + b)/2$ when $a = 0.5$, $b = 0.65$, $T_a = 315$ and $T_b = 310$.

13. The velocity v of molecules in the Cartesian coordinate system is represented via its coordinate components v_x, v_y and v_z by the equation

$$v = \sqrt{v_x^2 + v_y^2 + v_z^2}$$

Calculate the velocity for v_x, v_y and $v_z = 2.1$, -3.1 and 0.72, respectively.

14. The decay process of radioactive substances with time is described by the equation

$$N = N_0 \left(\frac{1}{2}\right)^t$$

where t is the number of half-life periods elapsed (the half-life being the time it takes for a decaying substance to disintegrate by half), N_0 is the initial amount of the substance and N is the residual amount.

(a) Calculate N for $t = 0$, 8, 16 and 24 days for substances with $N_0 = 500$ µCi (microCuries).

(b) Display the data as a two-column table, the first column t and the second N.

15. The Fourier series is one of the best instruments for description of complex biological shapes; the first terms in these series can be obtained as

$$F = a_1\cos(x + \psi_1) + b_1\sin(x + \psi_2)$$

Calculate F for coefficients a_1 and b_1 of 1.2 and 0.7, respectively, phase angles ψ_1 and ψ_2 of 0.1 and 0.2, respectively, and $x = \pi/7$.

16. The molecular weights (g/mol) for ten amino acids are given in the table

Ala	Arg	Asn	Asp	Cys	Gln	Glu	Gly	His	Ile
89	174	132	133	121	146	147	75	155	131

(a) Generate a Name vector with acid name notations;
(b) Generate a Weight vector with molar weight values;
(c) Generate a 10 × 2 matrix in which the first column are the Names and the second the Weights; use the num2str command for string representation of values in the matrix columns;
(d) Display the matrix with the disp command.

17. Use the table in Exercise 16 to list those amino acids with molecular weights (a) less than 100, (b) between 100 and 150, and (c) more than 150. Display the results in such a form that the title appears for each

group and then each row shows the amino acid name and molecular weight; use the disp commands.

18. The wind chill index can be calculated by the expression

$$w = -(35.75 - 0.4275T)v^{0.16} + 0.6215T + 35.74$$

in which the wind velocity v (mph) takes values from 10 to 60 in steps of 10 mph and the air temperature T (F) from 40 to –40 in steps of 20°F.

(a) Calculate w for each of the given values of v and T using the for ... end statements;

(b) Display the results with the fprintf command showing only integer values (digits preceding the decimal point only).

19. Calculate the wind chill index using the above expression without for ... end statements, using vectors only. Note that the first term on the right-hand side of the equation for w represents a vector product and should generate a matrix – it cannot be summed with the vector represented by the second term as the vector and matrix differ in size and should be aligned; use the ones command.

20. In a laboratory the number of bacterial cells was measured every hour from 0 up to 7 hours; the results were 100, 200, 400, 800, 1600, 3200, 6400 and 12 800.

(a) Generate a vector t with the elapsed time data;

(b) Generate a vector N with the numbers of bacterial cells;

(c) Generate and display an 8 × 2 matrix in which the first row represent t and the second N;

(d) Determine the minimal and maximal values for the bacterial growth data;

(e) Determine the integer mean value for the bacterial cells data, using the round command;

(f) Calculate the range r as the difference between the maximal and minimal values of the number of bacterial cells;

(e) Display all the results with a single fprintf command, so that each value appears on a new line with its nomenclature (e.g. Average = 3187).

21. Weather conditions have a strong influence on mosquito populations. For example, an experiment has shown that at monthly rainfall q of 2, 7, 11, 18 and 23 mm the mosquito density d (average monthly number

of mosquitoes trapped by all mosquito monitoring stations) was 0.8, 2.1, 4, 4.9 and 5.7, respectively. These data can be described by the linear equation $q = a_1 + a_2 d$ in which the coefficients a_1 and a_2 are obtainable from the following set of equations

$$a_1 n + a_2 \sum_{i=1}^{n} d_i = \sum_{i=1}^{n} q_i$$

$$a_1 \sum_{i=1}^{n} d_i + a_2 \sum_{i=1}^{n} d_i^2 = \sum_{i=1}^{n} q_i d_i$$

where n is the number of observed values.

(a) Find the coefficients solving the set with the left division rule;

(b) Find the coefficients with the right division rule;

(c) Show the results in the form of a linear equation with the coefficient values obtained displayed to three decimal places.

2.5 Answers to selected exercises

2. 3.141592653589793
3. $y = \log 10(10)/\log(2)$
6. `randi`
8. T'
11. $\psi = 7.5962$
13. $v = 3.8129$
15. $F = 1.3889$
18.

   Wind chill index

34	9	−16	−41	−66
30	4	−22	−48	−74
28	1	−26	−53	−80
27	−1	−29	−57	−84
26	−3	−31	−60	−88
25	−4	−33	−62	−91

19. Same answer as in Exercise 18.
21. The equation is $q = 0.648 + 0.234 * d$.

Published by Woodhead Publishing Limited

3

MATLAB® graphics

Plots are one of the best and most visually accessible means for presenting information in general and results of calculations in particular, especially in science and engineering. MATLAB® has a wide variety of commands for graphics. Available commands allow us to design various types of two- (in other words, XY or 2D) and three-dimensional (XYZ or 3D) plots.

2D graphics include plots with linear, semilogarithmic or logarithmic axes, bars, histograms, pies, polar and multiple plots, and many more. A plot can comprise one or several curves or surfaces, and several graphs can be plotted in the same Figure Window. Formatting can regulate the desired line style or marker type, its thickness and color, and add a grid, text or legend.

Plots having three axes are frequently used to represent data involving more than two variables. MATLAB® provides a variety of features for visualization of 3D data. These include spatial line plots, mesh and surface plots, geometrical figures and images. Generated plots can be formatted by commands or interactively from the Figure Window.

The most important commands for 2D and 3D plotting are described below. It is assumed that the reader has mastered the preceding chapter, and therefore lesser commands have inline explanations, and are written not within special frames (as in the second chapter), but as inline MATLAB® comments, next to the percentage sign (%).

3.1 Generation of XY plots

The basic command used for XY plotting is `plot`, which in its simplest form can be written as:

```
plot(x,y)    or    plot(y)
```

where x and y are two vectors of equal length, the first being used for horizontal and the second for vertical axes.

In the second plot-form the y values are plotted versus their indices.

After inputting the plot command with given values of x and y, the curve y(x) is created in the MATLAB® Figure Window with a linear scale by default. For example, we have mass (g/L) – time (min) data for an aerobic biomass process which are presented in Table 2.2. Let us plot a graph in which the x-axis has time values and the y-axis has biomass values. To create the plot we have to type in the Command Window:

```
>> t=[0:25:150 200];
>> b_mass=[5.15 5.21 5.5 6.55 7.15 7.75 7.59 7.45];
>> plot(t,b_mass)
```

Entering these commands in the Figure Window, the biomass–time plot resembles that shown in Figure 3.1.

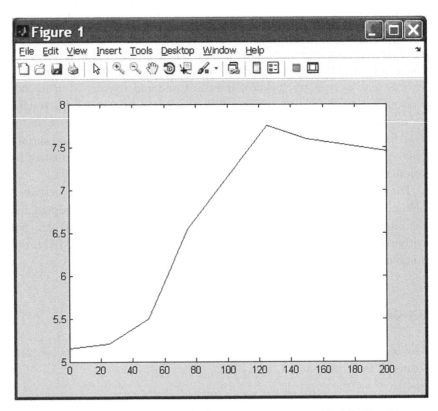

Figure 3.1 Biomass data plotted in the Figure Window with default settings.

Published by Woodhead Publishing Limited

The line style and marker type, its thickness and color can be regulated using the `plot` command with additional optional arguments:

`plot (x,y, 'Line Specifiers', 'Property Name', 'Prperty Value')`

where Line Specifiers determines the line type, the marker symbol and the color of the plotted lines (see Table 3.1); Property Name assigns properties to the specified Property Values and can be taken from Table 3.2.

Table 3.1 Line style, color and marker type specifiers*

Style, color or type	Specifier	Style, color or type	Specifier
Solid (default)	–	Magenta	m
Dotted	:	Circle	o
Dash-dot	–.	Asterisk	*
Dashed	– –	Point	.
(none)	no line	Square	s
Blue (default for single line)	b	Diamond	d
Green	g	Plus	+
Red	r	Triangle (inverted)	v
Black	k	Triangle (upright)	^
Yellow	y	Five-pointed asterisk	p
Cyan	c	Six-pointed asterisk	h

*Incomplete.

Table 3.2 Property names and property values*

Property name	Purpose	Property value
`LineWidth` or `linewidth`	Specifies the width of the line	A number in points (1 point = 1/72 inch). The default width is 0.5
`MarkerSize` or `markersize`	Specifies the size of the marker (by the '.' symbol)	A number in points
`MarkerEdgeColor` or `markeredgecolor`	Specifies the color of the marker or the edge color for filled markers	Color in accordance with specifiers in Table 3.1
`MarkerFaceColor` or `markerfacecolor`	The fill color for markers that are closed shapes	Color in accordance with specifiers in Table 3.1

*Incomplete.

Published by Woodhead Publishing Limited

Line specifiers, property names and property values are typed in the `plot` commands as strings in quotes (inverted commas). The specifiers and property names with their values can be written in any order, and one or more of them can be omitted. The omitted properties would be taken by default.

Some examples of application of specifiers and properties:

`plot(y,'-c')`   plots the cyan solid line with *x*-equidistant *y* points.

`plot(x,y,'g')`   plots the greed solid (default) line that connects the points.

`plot(x,y,'--p')`   plots the blue (default) dashed line, points marked with the five-pointed asterisks.

`plot(x,y,'k:h')`   plots the black dotted line, points marked with the six-pointed asterisks.

```
plot(t,b_mass, ...
'-mo','linewidth',3,'markersize',10,'MarkerEdgeColor',
...'k','MarkerFaceColor','y')
```

plots the magenta three-point solid line (1 point has size 1/72 inch), and *x,y*-values marked with the 10-point black-edged yellow circles.

Entering the last command with previously used biomass–time data, we get the plot shown in Figure 3.2.

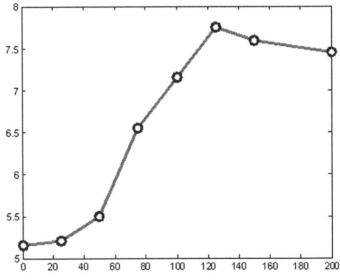

Figure 3.2 Biomass data generated with specifiers and property settings in the `plot` command.

Published by Woodhead Publishing Limited

To close one Figure Window type and enter the `close` command in the Command Window; to close more than one Figure Window, use the `close all` command.

3.1.1 Two or more curves on a single 2D plot

To plot two or more curves in the same graph, at least two options can be used in MATLAB®: the first by typing the pairs of x,y-vectors into the `plot` command and the second by using the `hold on` ... `hold off` commands.

3.1.1.1 The `plot` command option

The form of the `plot` command to create two or three curves in a single plot is:

> `plot(x1,y1,x2,y2)` or `plot(x1,y1,x2,y2,x3,y3)`

These commands create graphs with two and three curves, respectively, where `x1` and `y1`, `x2` and `y2`, `x3` and `y3` are the pairs of equal-length vectors containing the x,y data. For more than three curves, new x,y-pairs should be added in the `plot` command. As an example, to plot two trigonometric functions, sine and cosine, in the same plot, we enter in the Command Window the following commands (without comments):

```
>> x=0:pi/100:3;          % create vector x with values between 0 and 3
                          with step % pi/100
>> y1=sin(x);y2=cos(x);   % create vectors y1 and y2
>> plot(x,y1,x,y2, '--k')  % create two lines in the same plot
```

The resulting plot is shown in Figure 3.3 on p. 54.

3.1.1.2 The `hold` command option

This command is suitable when one plot already exists and we wish to add a new curve to it. In this case the `hold on` command must be typed and the new curve created by entering a new `plot` command. To complete the `hold on` process, we enter the `hold off` command that stops the hold process and shows the next graph in the new Figure Window.

As an example, the function $\sin(x^2)$ can be added to an existing graph on Figure 3.3 by entering the additional commands:

Published by Woodhead Publishing Limited

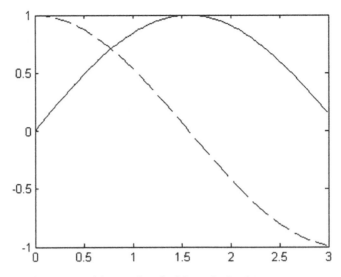

Figure 3.3   Two curves (sine and cosine) in a single plot.

```
>> y=sin(x.^2);        % create new vector y
>> hold on
>> plot(x,y, ':r')     % add vector y to the same plot
>> hold off
```

The resulting plot is shown in Figure 3.4.

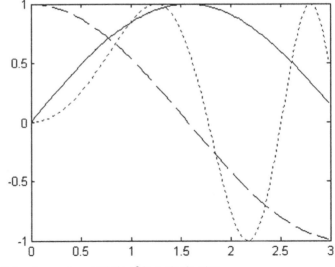

Figure 3.4   sin x, cos x and sin x^2 in a single plot.

3.1.1.3 Generation of several graphs on the same page

It is often necessary to place several plots on one page or, in other words, to multiply plots on the same page. For this, the subplot command has to be used, in the form:

subplot(m,n,p)

where m and n are the rows and columns of the panes into which the page is divided; p is the plot number that this command makes current (see Figure 3.5).

For instance:

subplot(2,2,4)   creates 4 panes ordered in 2 rows and 2 columns and makes subplot 4 current

subplot(2,3,2)   creates 6 panes ordered in 2 rows and 3 columns and makes subplot 2 current

subplot(2,1,2)   creates 2 panes in the same column and makes the last subplot current

subplot(1,2,1)   creates 2 panes in the same row and makes the first subplot current.

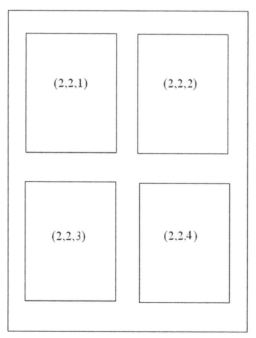

Figure 3.5 Possible arrangements of one page in four panes.

Published by Woodhead Publishing Limited

As an example, to generate four plots on one page: a one-point plot, sine plot, cosine plot and a circle in parametric form $x = \sin(t)$ and $y = \cos(t)$. The MATLAB® command to arrange the plots in this way are:

```
>> subplot(2,2,1)      % makes pane 1 current
>> plot(0.1, 'p',      % plots a five-point asterisk with a size of 10
'MarkerSize',10)
>> x=0:pi/100:2*pi;
>> subplot(2,2,2)      % makes pane 2 current
>> plot(x,sin(x))      % plots the sine curve
>> subplot(2,2,3)      % makes pane 3 current
>> plot(x,cos(x))      % plots the cosine curve
>> t=x;
>> subplot(2,2,4)      % makes pane 4 current
>> plot(sin(t),cos(t)); % plots the circle
```

The resulting plot is shown in Figure 3.6.

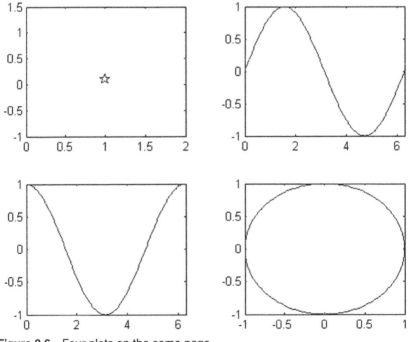

Figure 3.6 Four plots on the same page.

The plots can be arranged asymmetrically so that one can be placed into two or more columns or rows; for example, `subplot(2,2,[3,4])` creates four panes on the page and spans the third and fourth panes by the third plot (the bottom of the current page).

3.1.2 Formatting of 2D plots

All plotting commands described above produce bare plots. In practice, a figure must have a title, grid, axis labels, suitable axes ranges and some text. The plot can be formatted by including the specifying commands in the program created or interactively by the plot editor in the Figure Window. The first method is preferable when we intend to use the written program repeatedly, and the second can be used when the figure created is intended to be saved, e.g. for demonstration.

3.1.2.1 Formatting of 2D plots with specify commands

The formatting commands must be entered after the `plot` command. Some of these commands are described below.

The `grid` command

The `grid` or `grid on` commands add a grid to the created plot. The `grid off` command removes the grid lines from the latticed plot. For example, typing `grid` in the Command Window immediately after the commands that produced Figure 3.3 will add the grid to the figure.

The `axis` command

Some possible forms are:

$$axis([xmin\ xmax\ ymin\ ymax])$$

$$axis\ equal$$

$$axis\ square$$

$$axis\ tight$$

$$axis\ off$$

The first command adjusts the axes to the limits written in the square brackets; the second sets the same scale for both axes, the third sets the axes region to be square, the fourth sets the axis limits to the range of the data to be performed on the plot, and the last removes the axis from the plot.

As an example, enter the sequence of commands:

```
>> x=0:pi/100:pi;          % pi is maximal value of x
>> y1=sin(x);y2=cos(x);
>> plot(x,y1,x,y2, '--k')   % creates the plot with maximal x-axis limits 3.5
>> axis tight              % sets x_max and y_max to pi and 1, respectively
```

Figure 3.7 will be produced by these commands.

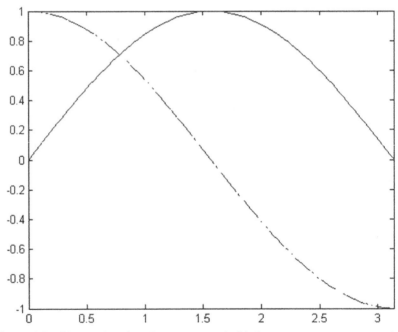

Figure 3.7 Sine and cosine plot constructed with the `axis tight` command.

The `xlabel`, `ylabel` and `title` commands

These commands provide text to each of the axes and at the top of the plot. The text must be written in string form. The commands have the form:

```
xlabel('text string')
ylabel('text string')
 title('text string')
```

There are provisions within the `text string` command to use Greek letters, font size and style, and also property options which can be written after the text for text angle, font name and color – see 'Formatting of text strings'.

The `text` and `gtext` commands

Existing forms are:

```
text(x,y,'text string')
 gtext('text string')
```

The first command directs the text at the point with coordinates *x* and *y*. The second command places it at the location chosen by the user. When this command is entered the Figure Window appears with two crossed lines and the user moves these lines by shifting the mouse to the point required, after which the text is entered by clicking on the mouse.

As an example, add the `title`, `xlabel`, `ylabel`, `text` and `grid` commands to the commands that constructed the plot in Figure 3.2:

```
>> title('Biomass vs Time')
>> xlabel('Time, min'),ylabel('Biomass,g/L')
>> text(75,6.3, 'Biomass')   % display word Biomass at point with x = 75
                               and y = 6.3
>> grid
```

The resulting plot is shown in Figure 3.8 on p. 60.

The `legend` command

This be written as:

```
legend('text string1','text string2',…,'Location',
               location)
```

The command shows the line type for each plotted curve and prints explanations written in 'text string1', 'text string2', … The 'Location' property is optional and specifies the location for the legend. Thus, for instance, location = 'NorthEastOutside' places the legend outside the axes on the right; location = 'Best' places the

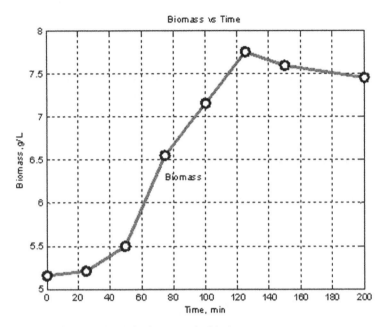

Figure 3.8 Biomass data plot formatted with the `xlabel`, `ylabel`, `title`, `text` and `grid` commands.

legend inside the axes at a better location (least conflict with the data in the plot). The default legend location is in the upper right-hand corner of the plot.

So, for Figure 3.7 inputting the command `legend('Sine','Cosine')` generates Figure 3.9 (see p. 61).

Formatting of text strings

The text strings in the described commands can be formatted by writing modifiers (special characters) inside the string or by including the options `PropertyName` with `PropertyValue` in the command after the text string.

Typing modifiers can be font name, style, size and color, Greek letters, or sub- and superscripts. Some useful modifiers are:

\b \it \rm	sets the type in bold, italic or normal, respectively.
_ ^	sets subscript and superscript, respectively; for example, `'^oC'` sets superscript for 'o' and the resulting text will display as °C.

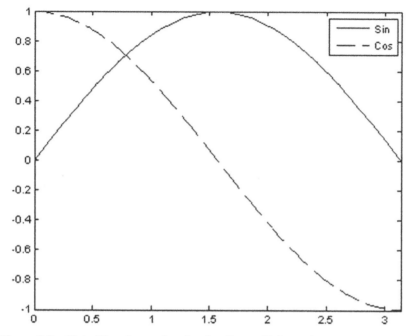

Figure 3.9   Plot of the sine and cosine functions with legend.

`\name of the Greek letter`	sets a Greek letter; for example, `\sigma` sets σ and `\Sigma` sets Σ.
`\fontsize{number}`	specifies the size of the type; for example, \fontsize{12} sets type at 12 point.
`\fontname{name}`	specifies the name of the font used; for example, \fontname{Arial} sets Arial font.

By including in the command the property (the name and the value), we can regulate the color of the text, or its background. Some property names and their values are:

`'Color','color specifiers from Table 3.1'`	sets the text color; for example, `'Color','b'` sets blue color for the text string.
`'BackgroundColor', 'color specifiers from Table 3.1'`	sets the background color (rectangular area); for example, `'BackgroundColor','y'` sets yellow color for background area.

Published by Woodhead Publishing Limited

A detailed explanation is available in MATLAB® Help and in the Text Properties section of the MATLAB® documentation.

3.1.2.2 Formatting of 2D plots with the plot editor

The Figure Window contains an assortment of formatting buttons and menu items. To start the plot edit mode the `Edit Plot` button, ⬚, must be clicked on the bar under the menu. The head of the Figure Window, with the menu and bar with most frequently used buttons, is shown in Figure 3.10. The properties of the axes and lines, and the whole figure, can be changed by means of the pop-up menu accessed by clicking the `Edit` Menu button. The title, axis labels, text and legend can be archived by means of the pop-up menu accessed by clicking the `Insert` button.

The text, legend and objects in the plot can be shifted by clicking on each of them. Double clicking of the curves, plotted points or axes activates the Property Editor, which allows many of the characteristics of the object selected to be changed or edited. Detailed information is available in the Help Window under the Editing Plot section.

3.2 Generation of XYZ plots

In MATLAB® there are three main groups of commands for lines, meshes and surfaces. These groups of commands together with various formatting commands are described below.

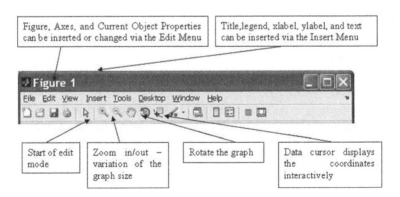

Figure 3.10 `Plot Editor` **buttons in the** `Figure Window`.

3.2.1 Generating lines in 3D plots

A line in 3D space connects points that are each described by three coordinates. Similar to 2D line plotting with the `plot` command, the `plot3` command is used to plot a 3D line. In its simplest form this command can be written

$$plot3(x,y,z),$$

or in a more complicated form

```
plot3(x,y,z,'Line Specifiers','Property Name','Property
                      Value')
```

Here x, y and z are the equivalent vectors with point coordinates, and the `Line Specifiers`, `Property Name` and `Property Value` have the same sense as in the 2D case.

In 3D plotting the `grid`, `xlabel` and `ylabel` commands and, by analogy, the `zlabel` commands are also used.

For example, a 3D plot can be produced as follows:

```
>> t=0:pi/100:6*pi;
>> x=t.*cos(2*t);
>> y=t.*sin(2*t);
>> z=t;
>> plot3(x,y,z, 'k', 'LineWidth',4)
>> grid
>> xlabel('x'),ylabel('y'),zlabel('z')
```

This algorithm computes the parametrically given coordinates $x = t \cdot \cos(2t)$, $y = t \cdot \sin(2t)$ and $z = t$ with t changed from 0 up to 2π in steps of $\pi/100$.

With the procedure completed the plot appears in the Figure Window and the line has the form shown in Figure 3.11 (see p. 64).

3.2.2 Meshes in 3D plots

To understand the main commands of the 3D plot, namely `mesh` and `surf`, it is useful to understand the mesh construction in MATLAB®. As every point in 3D space has three coordinates x, y and z, these must be given in reconstructing a surface. In other words, we must have two 2D matrices for the x- and y-coordinates, respectively, and calculate the matrix of

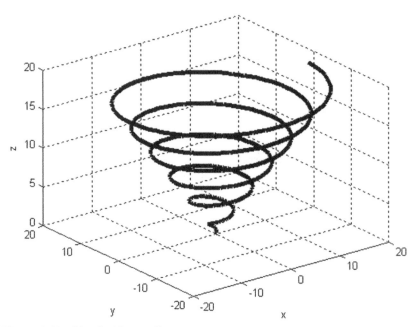

Figure 3.11   Line in 3D coordinates.

z-coordinates for every pair. The area of x and y coordinates for which the z-coordinates must be obtained is called the domain. Figure 3.12 overleaf shows an example of point representation in 3D space. The grid in the x–y plane is the domain given by the vectors x = –2…2 and y = –2…2. Each node in the x–y plane has a pair of x,y values. Writing all x-values ordered by rows, we obtain the X-matrix, and the same procedure yields the Y-matrix. For the case in Figure 3.12 matrices X and Y read:

$$X = \begin{vmatrix} -2 & -1 & 0 & 1 & 2 \\ -2 & -1 & 0 & 1 & 2 \\ -2 & -1 & 0 & 1 & 2 \\ -2 & -1 & 0 & 1 & 2 \\ -2 & -1 & 0 & 1 & 2 \end{vmatrix} \qquad Y = \begin{vmatrix} -2 & -2 & -2 & -2 & -2 \\ -1 & -1 & -1 & -1 & -1 \\ 0 & 0 & 0 & 0 & 0 \\ 1 & 1 & 1 & 1 & 1 \\ 2 & 2 & 2 & 2 & 2 \end{vmatrix}$$

With X and Y matrices available the z-coordinates must be obtained for every grid point using element-by-element calculation. The whole surface can then be plotted.

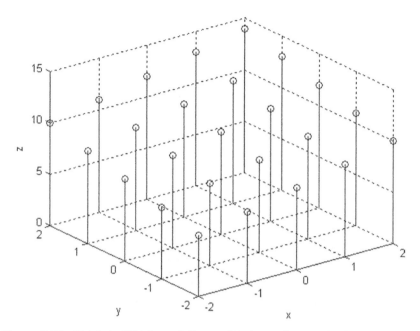

Figure 3.12 Points in 3D interpretation and their *x,y*-plane projection.

MATLAB® has a special command called meshgrid which allows us to calculate X and Y from the given vectors of *x* and *y*. The command has the form:

$$[X,Y]= \text{meshgrid}(x,y)$$

Here X and Y are the matrices of grid coordinates calculated in this function on the basis of the given *x* and *y* vectors that determine the domain dividing. When the *x* and *y* vectors are equal, this command may be simplified to

$$[X,Y]= \text{meshgrid}(x)$$

For example, the X and Y matrices can be defined for the case in Figure 3.12 as follows:

>> x=–2:2;
>> [X,Y]=meshgrid(x)

X =

-2	-1	0	1	2
-2	-1	0	1	2
-2	-1	0	1	2
-2	-1	0	1	2
-2	-1	0	1	2

Y =

-2	-2	-2	-2	-2
-1	-1	-1	-1	-1
0	0	0	0	0
1	1	1	1	1
2	2	2	2	2

The Z matrix can be calculated for instance via the expression

$$Z = X + Y + XY$$

which corresponds to the Z-coordinates of the points in Figure 3.12.

We can now generate a graph with the aid of the `mesh` command. This command has the simplest form:

$$\text{mesh}(X, Y, Z)$$

where X, Y and Z are the same matrices as in the example. This command also enables the mesh lines to be colored in the plot.

Thus, the program that plots the mesh surface reads:

```
>>x=-2:0.1:2;
>>[X,Y]=meshgrid(x);
>>Z=X+Y+Y.*X;
>>mesh(X,Y,Z)
>>xlabel('x'),ylabel('y'),zlabel('z')
>>grid on
```

The resulting plot is shown in Figure 3.13.

3.2.3 Surfaces in 3D plots

If the plot needs colored surfaces between the mesh lines, the `surf` command can be used. The form of this command is:

$$\text{surf}(X, Y, Z)$$

X, Y and Z being the same matrices as in the mesh command.

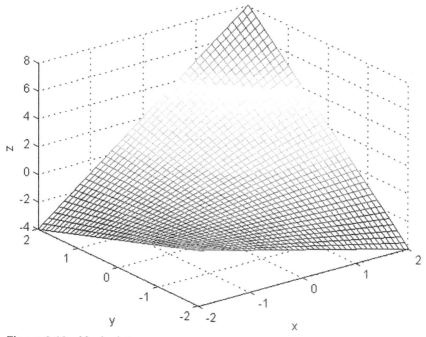

Figure 3.13 Mesh plot.

Using this command for the function in the previous example we can enter commands as follows:

```
>>x=-2:0.1:2;
>>[X,Y]=meshgrid(x);
>>Z=X+Y+Y.*X;
>>surf(X,Y,Z)
>> xlabel('x'),ylabel('y'),zlabel('z')
>>grid on
```

The resulting plot is shown in Figure 3.14.

The `surf` command and the `mesh` command can be used in the form `surf(Z)` or `mesh(Z)`. In this case the Z values are plotted versus the indices.

3.2.4 Formatting and rotation of 3D plots

Many of the 2D commands described in section 3.1.3 are suitable for 3D plot formatting, such as `grid`, `title`, `xlabel`, `ylabel` and `axis`. Many

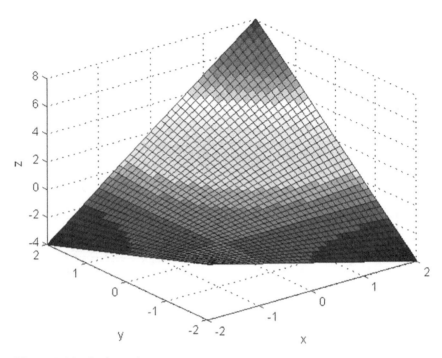

Figure 3.14 Surface plot.

additional commands exist to format a 3D plot. Some of these are described below.

3.2.4.1 The box command

The box on command draws a box around the plot, and on entering the box off command the drawn box disappears. Figure 3.15 illustrates use of the command for the previous example:

>>box on

3.2.4.2 The colormap command

Color plays a more important role in 3D plotting than in 2D plotting. Default color is automatically generated with mesh or surf according to the z-values. With the colormap command the colors can be set at a constant value. The command has the form

colormap(c)

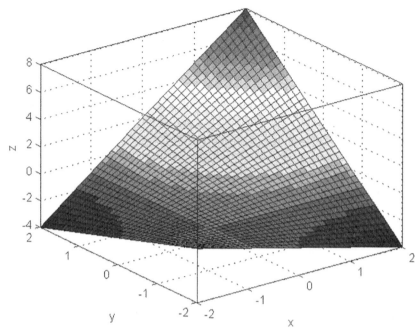

Figure 3.15 Boxed surface plot.

where c is a three-element vector in which the first element specifies the red color intensity, the second the green color intensity and the third the blue intensity (RGB); intensities are graded from 0 to 1, for example:

c=[0 0 0]   black
c=[1 1 1]   white
c=[1 0 0]   red
c=[0 1 0]   green
c=[0 0 1]   blue
c=[1 1 0]   yellow
c=[1 0 1]   magenta
c=[0.5 0.5 0.5]   gray.

If colormap([0 0 1]) is added to the commands that produced the mesh plot in Figure 3.14, we obtain this figure with the mesh lines changed in color to blue.

MATLAB® has some built-in color regulation commands in the form

colormap name

where the name can be jet, cool, winter, spring and a few others. For example, the colormap spring changes colors to shades of magenta and yellow.

3.2.4.3 *The* view *command*

The plot orientation relative to the viewer look can be regulated by the view command, having the form

```
view(az,el)
```

where az and el are azimuth and elevation angles: the first is the horizontal (*x,y*-plane) angle relative to the negative direction of the *x*-axes; the second is the vertical angle defined as the elevation from the *x,y*-plane. A positive az value is defined in the counter-clockwise direction. A positive el value corresponds to opening the angle in the direction of the *z*-axes. Both angles must be in degrees; default values are az = –37.5° and el = 30°.

The azimuth and elevation angles are shown on Figure 3.16.

The various planes can be viewed at the chosen appropriate angle, as follows:

- the *x,y*-projection of the 3D plot can be obtained with az = 0 and el = 90 – top view;
- the *x,z*-projection of the 3D plot can be obtained with az = el = 0 – side view [can be entered simply as view(2)];
- the *y,z*-projection of the 3D plot can be obtained with az = 90 and el = 0 – side view.

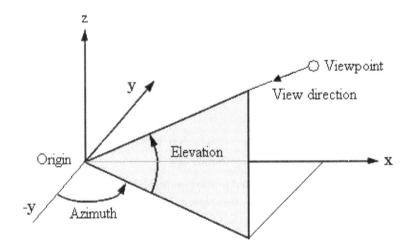

Default angles: Azimuth, Az=-37.5°, Elevation, El = 30°

Figure 3.16 Viewpoint, azimuth and elevation.

Published by Woodhead Publishing Limited

For example, the function $z = e^{-x^2 - y^2}$ can be plotted for default viewing angles, angles $az = 20°$ and $el = 35°$, and for the top and side views by the commands:

```
>> x=-3:.25:3;
>> [X,Y]=meshgrid(x);
>> Z=X.*exp(-X.^2-Y.^2);
>> subplot(2,2,1),surf(X,Y,Z)      % az = -37.5 and el = 30°
>> subplot(2,2,2),surf(X,Y,Z)
>> view(20,35)                     % az = 20° and el = 35°
>> subplot(2,2,3),surf(X,Y,Z)
>> view(2)                         % az = 0° and el = 90° - top view
>> subplot(2,2,4),surf(X,Y,Z)
>> view(0,0)                       % az = 0° and el = 0° - side view
```

which gives the result shown in Figure 3.17.

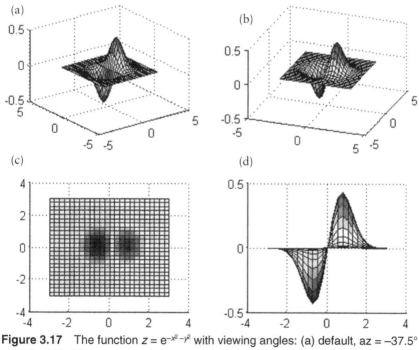

Figure 3.17 The function $z = e^{-x^2 - y^2}$ with viewing angles: (a) default, az = –37.5° and el = 30°; (b) az = 20° and el = 35°; (c) az = 0° and el = 90° (top view); (d) az = el = 0° (side view).

3.2.4.4 *The* rotate3d *command*

This command has the forms

rotate3d on

rotate3d off

Typing the rotate3d on or pressing 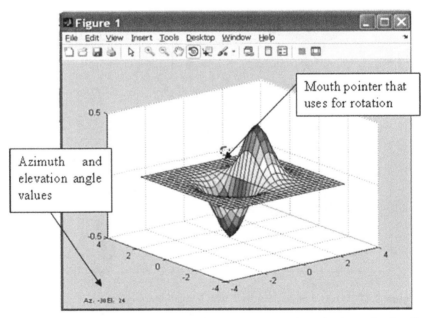 on the figure toolbar enables us to rotate the plot with the mouth pointer (see Figure 3.18), and to view the azimuth and elevation angles appearing in the bottom left-hand corner of the figure. For this, after entering the command in the Command Window, we should:

- go to the Figure Window,
- and by holding down the mouse button, rotate the plot and simultaneously view the az and el values that appear.

A view of the Figure Window in this regime is shown in Figure 3.18.

Plot rotation ends when the rotate3d off command is entered in the Command Window.

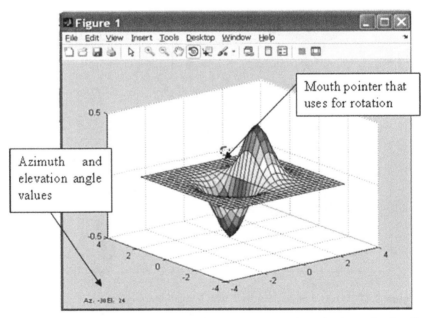

Figure 3.18 Plot in rotation regime with the rotate cursor and values of azimuth and elevation angles.

3.3 Specialized 2D and 3D plots

There are accessories to the commands of 2D and 3D graphics with the aid of which various specialized plots can be constructed. Some of them, such as errorbar, hist, semilogx and semilogy, are described below.

3.3.1 The errorbar command

Usually $y = f(x)$ data have some uncertainty or an error in the y-value and it is desirable to set limits to it in each data point. The errorbar command makes it possible to plot the points with error limits. The simplest form of this command is

```
errorbar(x,y,e)
```

where x and y are the data vectors and e is that of the symmetric (two sides equal) error.

For example, if the biomass–time data used in section 3.1 have a symmetric error of ±0.15 g/l at each point, the commands

```
>> t=[0:25:150 200];              % vector with the time data
>> b_mass=[5.15 5.21 5.5
6.55 7.15 7.75 7.59 7.45];        % vector with the biomass data
>> e = 0.15+zeros(1,length(b_mass));  % creates vector of the error values
>> errorbar(t,b_mass,e)           % plots graph with error bars
>> xlabel('Time, min'),ylabel('Biomass, g/l')
>> grid
```

yield, on entering, the plot shown in Figure 3.19.

A line style specifier may be included into the errorbar command for setting marker and/or line color and style; for example, in the previous plot the errorbar(t,b_mass,'--o') command changes the line to a dashed one and signs the data points with circles.

3.3.2 The hist command

The histogram is one of the most popular forms of data representation. Data values in a histogram are divided into intervals (bins) and plotted in the form of vertical bars, the height of which represents the number of data

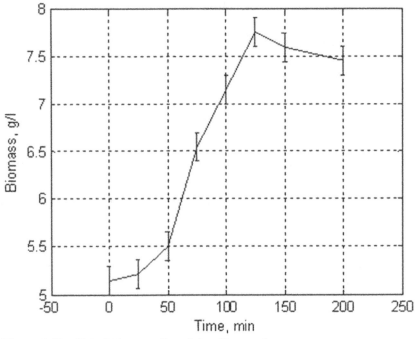

Figure 3.19 Plot of biomass–time data with error bars.

in each of them. Such a plot can be plotted with the aid of the hist command, whose simplest form is

hist(y) or n=hist(y)

where y is the vector containing the data points and n is the vector containing the number of data points in each of the 10 (default) equally spaced bins.

The first command yields the histogram plot, while the second yields only numerical output.

For example, the following data points are the weights (in mg) of 33 vials from a lot: 65, 80, 95, 93, 67, 81, 90, 93, 92, 83, 86, 83, 90, 94, 93, 96, 96, 100, 96, 50, 75, 81, 65, 88, 81, 60, 57, 61, 68, 77, 75, 76 and 70. A histogram can be plotted with the commands

>> y=[65 80 95 93 67 81 90 93 92 83 86 83 90 94 93 96 96 100 96 50 ...
75 81 65 88 81 60 57 61 68 77 75 76 70];
>> hist(y)
>> xlabel('Weight,mg'),ylabel('Number of weights per one bin')

The resulting plot is shown in Figure 3.20.

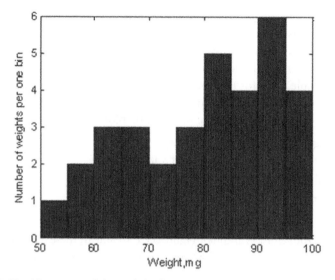

Figure 3.20 Histogram of the weight data.

By using the n=hist(y) command in this example, the following frequencies will be displayed in the Command Window:

```
>> n=hist(y)
n =
    1    2    3    3    2    3    5    4    6    4
```

The hist command has additional forms that allow us to set the required number of bins or output the values of the bin locations; detailed information is obtainable by entering help hist in the Command Window.

3.3.3 Plots with semi-logarithmic axis

In bioscience and engineering, coordinates with one axis on a logarithmic scale frequently have to be used. Log scales allow us to present values in a wider range than can be done with linear axes and can be used to plot the various exponential-like relationships that are widespread in biotechnology, such as expressions for first-, second- or higher-order reactions, population growth rates, radioactive decay and sterilization. For this, the semilogy or semilogx commands should to be used. In their simplest form:

```
semilogy(x,y,'LineSpec'),    semilogx(x,y,'LineSpec')
```

Published by Woodhead Publishing Limited

The `semilogy(x,y)` command generates a plot with a log-scaled (base 10) y-axis and a linear x-axis; `semilogx(x,y)` generates a plot with a log-scaled x-axis and a linear y-axis; `LineSpec` (optionally) specifies the line type, plot symbol and color for the lines drawn in the semi-log plot.

For example, data for the decay of a radioactive substance are: 400, 200, 100, 50 and 25 disintegrations per minute at 0, 1, 2, 3 and 4 hours, respectively. A semi-logarithmic plot can be plotted with the commands:

```
>> N=[400 200 100 50 25]; t=0:4;
% plot N on a semi-log y-axis and a linear x-axis
>> semilogy(t,N, '-o')
>> xlabel('Time, h'), ylabel('Disintegrations/minute')
>> grid
```

The data in the semi-log graph of Figure 3.21 are linear; these data on a normal graph produced by the `plot` command generate an exponential curve.

3.3.4 Additional commands for 2D and 3D graphics

Table 3.3 lists additional commands for 2D and 3D plotting; it gives the format of the corresponding basic command with short explanations, examples and the resulting plots. A complete list of 2D, 3D and specialized plotting functions can be obtained by entering the following commands: `help graph2d`, `help graph3d` or `help specgraph`.

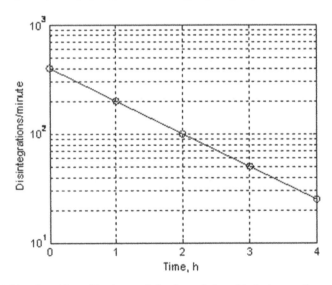

Figure 3.21 Semi-logarithmic graph for the relationship between time elapsed and residual radioactivity.

Published by Woodhead Publishing Limited

Table 3.3 Additional commands and plots for 2D and 3D graphics*

Commands	Examples	Plots
`figure(h)` creates a new Figure Window with number *h*	>>figure(2)	
`fplot('function', limits)` plots a function $y = f(x)$ with specified *x*-limits (the limits of the *y*-axis may be added)	>>fplot ('exp(.1*x)', [10,60])	
`polar(theta,rho)` generates a plot in polar coordinates in which `theta` and `rho` are the angle and radius, respectively.	>>th=linspace (0,2* pi,150); >> r=4*cos(3*th); >> polar(th,r)	
`loglog(x,y)` generates plot with log-scaled (base 10) *x*- and *y*-axes	>> x=linspace (0.1,20,100); >> y=5+exp (-0.5*x); >> loglog(x,y)	
`sphere` or `sphere(n)` plots a sphere with 20 or n mesh cells, respectively	>> sphere(40)	
`cylinder` or `cylinder(r)` draws an ordinary and a profiled cylinder with the profile given by the expression for *r*	>> t =0:pi/10:2*pi; >> r=atan(t); >> cylinder(r)	

Table 3.3 *Continued*

contour(X,Y,Z,v) displays in the *x,y*-plane isolines of matrix Z; Z is interpreted as the height with respect to the *x,y*-plane, v is the number of contour lines, or the vector-specified contour lines. The form c=contour(X,Y,Z,v) with clabel(c) displays level values c of the isolines	>> x=-2:.2:2; >> [X,Y]=meshgrid (x); >> Z=X.*exp (-X.^2-Y.^2); >> contour (X,Y,Z,7); >>% Or with level values >> c = contour (X,Y,Z,7); >> clabel(c)	
contour3(X,Y,Z,n) displays in the *x,y*-plane isolines of matrix Z; Z is interpreted as the height with respect to the *x,y*-plane, and n as the number of contour lines	>> x=-2:.2:2; >> [X,Y] = meshgrid(x); >> Z=X.*exp (-X.^2-Y.^2); >> contour3 (X,Y,Z,7);	
surfc(X,Y,Z) generates surface and contour plots together	>> x=-2:.2:2; >> [X,Y]=meshgrid(x); >> Z=X.*exp (-X.^2-Y.^2); >> surfc(Z);	
bar(x,y) displays the values in a vector or matrix as vertical bars	>> x=55:5:100; >> y=[1,2,1,0,1, 4,3,3,5,5]; >> bar(x,y) >> xlabel ('Grades'), >> ylabel ('Number of Grades per Bin')	
bar3(Y) generates 3D bar plot data grouped into columns	>> Y=[2 3.4 7.1 2.2 1; 3 6 4.1 3 2; 0.3 5 6.3 4 2]; >> bar3(Y)	

Published by Woodhead Publishing Limited

Table 3.3 *Continued*

stem(x, y) displays data as lines extending from a baseline along the x-axis. A circle (default) terminates each stem	>> x=55:5:100; >> y=[1,2,1,0,1, 4,3,3,5,5]; >> stem(x,y) >> xlabel ('Grades'), >> ylabel ('Number of Grades per Bin')	
pie(x) draws a pie chart using the data in x. Each element in x is represented as a slice.	>> x=[56 68 42 91 100]; >> pie(x) >> title ('Group Grades')	
pie3(x,explode) generates a pie chart; explode specifies an offset of a slice from the center of the chart; explode is a vector of the same length as x, in which 1 denotes an offsetted slice and 0 a plain slice	>> x=[56 68 42 91 100]; >> explode=[0 1 0 0 0]; >> pie3 (x,explode) >> title('Group Grades')	

*The commands are described in their simplest form.

3.4 Application examples

3.4.1 Two species concentration change in an opposed reaction

Consider that for two species A⇌B with initial concentrations $[A]_0$ and $[B]_0$ equal 0.25 and 0 mol/L, respectively, the concentrations at time $t = 0$, 0.1, ..., 1 can be determined from the following equations:

$$[A] = [A]_0 \frac{1}{k_f + k_b}\left(k_b + k_f e^{-\left(k_f + k_b\right)t}\right)$$

$$[B] = [A]_0 \frac{k_f}{k_f + k_b}\left(1 - e^{-\left(k_f + k_b\right)t}\right)$$

The reaction rate constants k_f and k_b equal 2 and 1 min^{-1}, respectively.

Problem: Calculate and plot the results, and present two forms of a concentration–time graph: (a) $[A] - t$ and $[B] - t$ curves on the same graph, and (b) each of these curves on different graphs but on the same page.

The commands for presentation (a) are

```
>>% determine the initial concentration and reaction rate constants
>> Ao=.25;kf=2;kb=1;
>> t=0:.1:1;                    % create time vector
>> A=Ao*1/(kf+kb)*             % calculate concentrations A
(kb+kf*exp(-(kf+kb)*t));
>> B=Ao* kf/(kf+kb)            % calculate concentrations B
*(1-exp(-(kf+kb)*t));
>>% plot A and B vs. T; A-solid line (default), B-dashed
>> plot(t,A,t,B, '--')
>>% this and the next commands are used to format the current plot
>> xlabel('Time, min'),
>>ylabel('Concentration,g/l')
>> grid
>> title('Two species concentration vs time')
>>legend('The species A', 'The species B')
```

The plot created by these commands is shown at the top of p. 81.

For the second representation, the plotting command in the preceding text should be changed to:

```
>> subplot(2,1,1)
>> plot(t,A)
>> title('Concentration of the A-spacies vs. Time')
>> xlabel('Time, min'),ylabel('Concentration,g/l')
>> grid
>> subplot(2,1,2)
```

Published by Woodhead Publishing Limited

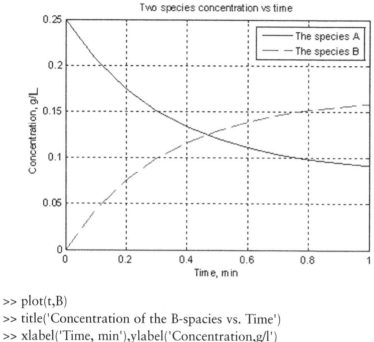

```
>> plot(t,B)
>> title('Concentration of the B-spacies vs. Time')
>> xlabel('Time, min'),ylabel('Concentration,g/l')
>>grid
```

After inputting, these commands produce two graphs on the same page:

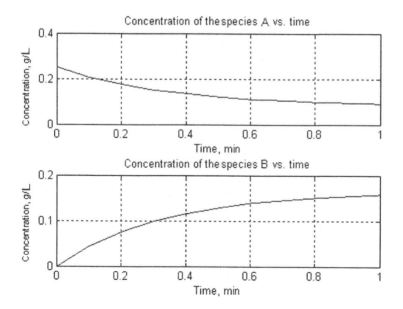

3.4.2 Microorganism growth curve

One type of bacterium has a doubling time of 30 minutes. The initial number N_0 of bacteria in a flask was 100. The general equation for bacterial population growth is

$$N = N_0 2^n$$

where n is the number of generations that have elapsed; in our case after 3 hours the number of elapsed generations n is $3 \times 60/30 = 6$.

<u>Problem:</u> Calculate and plot bacterial growth over 3 hours in two plots on one page: (a) N versus time t, and (b) log N versus t; mark calculation points with a circle (the letter 'o').

The commands used to solve this problem are:

```
>> No=100;t_double=.5;
>> n=0:6;                % Number of generations elapsed
>> N=No*2.^n;
>> t=n* t_double;        % Time elapsed
>> subplot(2,1,1)
>> plot(t,N, '-o')
>> title('Bacterial growth during 3 hours')
>> xlabel('Time elapsed, hours'), ylabel('Number of bacteria, cell')
>> grid
>> subplot(2,1,2)
>> semilogy(t, N, '-o')   % the command for a semi-logarithmic plot
>> title('log plot of bacterial growth during 3 hours')
>> xlabel('Time elapsed, hours'), ylabel('Log of number of cells')
>> grid
```

The figure produced is shown on p. 83.

3.4.3 Air density at atmospheric pressure

Under the ideal-gas approach, the density, ρ, of air at atmospheric pressure, p, and different temperatures, T, can be calculated with the ideal gas equation of state as:

$$\rho = p/RT$$

with $p = 101.3$ kN/m³, $R = 0.286$ kJ/(kg°K) and T in degrees Kelvin.

Published by Woodhead Publishing Limited

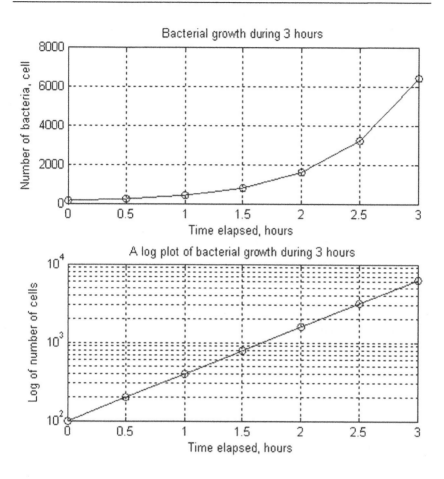

The experimental air density values measured with mean two-sided error $\delta\rho = 0.1\%$ at 0–100 °C (or 273.15–373.15 °K) are given in Table 3.4.

<u>Problem</u>: Calculate and plot the air density as determined theoretically (via the equation of state) and experimentally (from measured data with data errors in each point).

The commands to calculate and plot air densities at atmospheric pressure and temperatures of 273.15–373.15 °K are:

Table 3.4 Air temperature–density data

T (°C)	0	20	40	60	80	100
ρ (kg/m³)	1.293	1.205	1.127	1.067	1	0.946

```
>> T=273.15:20:373.15;              % in °K
>> ro_exp=[1.293 1.205 1.127 1.067 1 0.946];     % in kg/m³
>> p=101.3;                         % atmospheric pressure in kN/m²
>> R=.286;                          % in kJ/(kg°K)
>> ro_theor=p./(R*T);
>> error=0.001* ro_exp;
>> errorbar(T, ro_exp, error,'.b'),hold on
>> % plots experimental points with error bars
>> plot(T, ro_theor), grid
>> % plots theoretically calculated curve
>> xlabel('Temperature, ^oK'),ylabel('Density, kg/m^3')
>> title('Air Density'),
>> legend('experimental points', 'theoretical curve'), hold off
```

The resulting plot is:

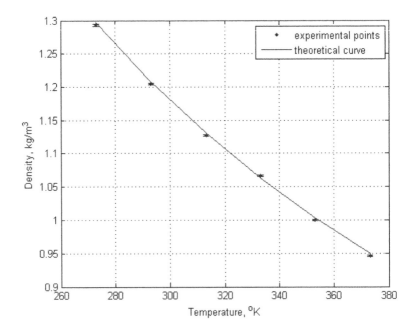

3.4.4 Elliptic Gaussian function

The 2D elliptic Gaussian distribution function for the uncorrelated variables x (e.g. species weight) and y (e.g. species volume) having a bivariate normal distribution is given by

$$f(x,y) = \frac{1}{2\pi\sigma_x\sigma_y} e^{-\left[\frac{(x-\mu_x)^2}{2\sigma_x^2} + \frac{(x-\mu_y)^2}{2\sigma_y^2}\right]}$$

where μ_x and μ_y are the mean values of x and y, and σ_x and σ_y are the standard deviations.

<u>Problem</u>: Calculate and plot the distribution function for $\mu_x = 90$, $\mu_y = 130$, $\sigma_x = 10$, $\sigma_y = 20$ in x-axis limits $\mu_x \pm 3\sigma_x$ and y-axis limits $\mu_y \pm 3\sigma_y$.
 The steps used to solve this problem are:

- define 40-point vectors x and y with the `linspace` command;
- create an X,Y grid in the ranges of the x, y vectors by using the `meshgrid` command;
- calculate f via the above expression;
- generate an X,Y,f plot with the `surf` command.

The commands are:

```
>> sigma1=10; sigma2=20;
>> mu1=90; mu2=130;
>> setting limits for x, y, sigma_x, sigma_y
>> xlim=3*sigma1; ylim=3*sigma2;
>> xmin=mu1-xlim;xmax=mu1+xlim;
>> ymin=mu2-ylim;ymax=mu2+ylim;
>> x=linspace(xmin,xmax,40);          % defining vector x
>> y=linspace(ymin,ymax,40);          % defining vector y
>> create X,Y grid from the x,y vectors
>> [X,Y]=meshgrid(x,y);
>> f=1/(2*pi*sigma1*sigma2)*exp(-((X-mu1).^2/(2*sigma1^2)+( ...
Y-mu2).^2/(2*sigma2.^2)));            % calculate element-wise the f
>> surf(X,Y,f)                        % plot surface graph
>> xlabel('x'),ylabel('x'),zlabel('f(x,y)')
```

The configuration generated by these commands is shown on p. 86:

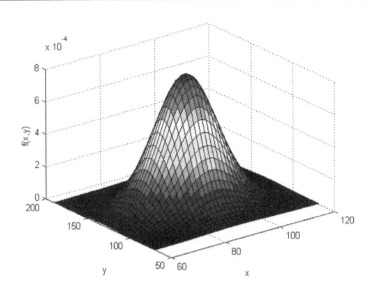

3.4.5 One-dimensional transient diffusion

The analytical solution of the 1D diffusion equation has the form

$$c = \frac{1}{\sqrt{4\pi Dt}} e^{-\frac{x^2}{4Dt}}$$

where c is a species concentration that changes with coordinate x and time t, and k is the diffusion coefficient.

Problem: Calculate and generate a 3D surface plot in which x and t are in the horizontal plane and the z-axis is the concentration; if $D = 1$ cm²/s, $x = 0, 0.1, ..., 2$ cm, and $t = 0.1, 0.2, ..., 2.1$ s, present the plot with azimuth and elevation angles of 123° and 26°, respectively.

The program follows these steps:

- create the vectors of x and t;
- create an X,Y grid in the ranges of the x and t vectors, respectively, by using the meshgrid command;
- calculate c for each pair of X and Y values by the above expression;
- generate a 3D plot by the X,Y,c-values determined;
- set required view point for the plot generated.

The commands to solve this problem are:

```
>> x=0:.1:2;t=0.1:0.2:2.1;                    %D=1 and is not used here
>> % creates the X,Y grid from the x,t vectors
>> [X,Y]=meshgrid(x,t);
>> % calculate element-wise the value of c
>> c=1./sqrt(4*pi*Y).*exp(-(X).^2./(4*Y));
>> surf(X,Y,c);                               % plot surface graph
>> xlabel('Distance');ylabel('Time');zlabel('Concentration')
>> view(123,26)                               % az = 123° el = 26°
```

3.4.6 First-order reaction: concentration changes with time and temperature

In the first-order reaction $A \rightarrow$ Product, the concentration $[A]$ changes with time t according to

$$[A] = [A]_0 e^{-kt}$$

where $[A]_0$ is the initial concentration of A.

The rate constant k changes with temperature according to the Arrhenius equation:

$$k = ae^{-E_a/RT}$$

in which a is a frequency factor, E_a is the activation energy, T is temperature and R is the gas constant.

The values of the parameters in these equations are $a = 10^{14}$ s^{-1}, $E_a = 77\,000$ J/mol, $T = 293, 294, \ldots, 303$ K, $R = 8.314$ J/(K mol), $[A]_0 = 0.25$ mol/L, and $t = 1, 1.2, \ldots, 4$ s.

Problem: Calculate and generate a 3D surface plot of the concentration (z-axis) as a function of time (x-axis) and temperature (y-axis). Present the plot with azimuth and elevation angles of 67° and 22°, respectively. Change colors to the autumn combination of colors.

The following steps should be used to solve this problem:

- determine a, E_a, R and $[A]_0$ and create the vectors for t and T;
- create an X,Y grid in the ranges of the t and T vectors, respectively, by using the meshgrid command;
- calculate k and $[A]_0$ for each pair of X and Y values by the above expression;
- generate a 3D plot using the defined values of X, Y and $[A]$;
- set the required view point and color combination.

The commands are:
```
>> ea=77e3;a=1e14;R=8.314;a0=.25;
>> t=1:.2:4; T=293:303;
>> [X,Y]=meshgrid(t,T);
>> k=a*exp(-ea./(R*Y));
>> A=a0*exp(-k.*X);
>> surf(X,Y,A)
>> xlabel('Time, sec'), ylabel('Temperature, K'),
>> zlabel ('Concentration, mol/L'),
>> view(67,22)
>> colormap spring
```

The resulting plot is shown at the top of p. 89.

3.5 Questions for self-checking and exercises

1. The plot3 command generates on a 3D graph: (a) a surface, (b) a surface mesh, (c) a line?
2. Which sign or word specifies a circled point on a plot: (a) +, (b) the word 'circle', (c) o, (d) t?

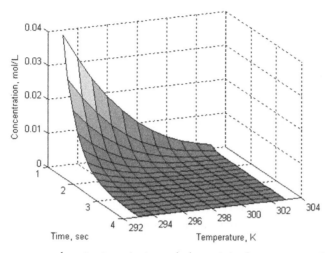

3. The command `subplot` is intended to: (a) place a new plot into a previously generated plot, (b) place several plots on the same page, (c) change colors on the graph?

4. A biotechnology company is using a bacterial broth to produce an antibiotic; the pH of the broth was measured for 5 hours every 30 minutes, giving: 6.12, 5.13, 5.84, 6.53, 6.12, 6.3, 6.04, 5.79, 5.94, 6.03, 6.12. Plot the data with a solid line between data points marked with a diamond; add axis labels, grid and a caption.

5. Generate three plots on one page for the following geometrical figures.

Astroid:

$$x = 4\cos^3 \varphi$$
$$y = 4\sin^3 \varphi$$

with $0 \leq \varphi \leq 6\pi$.

Archimedes' spiral

$$x = 2\varphi \cos^3 \varphi$$
$$y = 2\varphi \sin^3 \varphi$$

with $0 \leq \varphi \leq 6\pi$.

Leminscate

$$x = \frac{\varphi \cos \varphi}{1 + \sin^2 \varphi}$$

$$y = \frac{\varphi \sin \varphi \cos \varphi}{1 + \sin^2 \varphi}$$

with $0 \leq \varphi \leq 2\pi$.

Place the astroid and Archimedes' spiral in the first graph lines and the lemniscate in the second line. Make the size of the current axis `square` for the astroid and Archimedes' spiral, and add a caption to each plot.

6. Plot the function

$$f_1(x) = 13x^7 - 4x^5 + 0.6x^3 - 2$$

for $0 \leq x \leq 4$.

Add to the graph the new function

$$f_2(x) = e^{1.3x}$$

with $1 \leq x \leq 5$.

Add axis labels, grid and legend to the graph.

7. The kinematic viscosity η (in cP) of a bio-oil as a function of temperature T (in K) is given by the Andrade-type expression

$$\eta = Ae^{\frac{B}{T}}$$

where $A = 0.0005806$ and $B = 3382$.

Measured viscosities at each temperature are given in the table below

T (K)	298.15	308.15	318.15	328.15
η (cP)	49.14	33.45	23.92	17.71

The uncertainty of the experimental data is ±0.5%.

Plot the calculated and measured viscosity values. Give the error bars at each measured point, add axis labels, a caption and legend.

8. Plot the surfaces of the following bodies on one page. Make the current axis `equal` in size and add a caption to each figure:

    (a) Ellipsoid

$$x = a \cdot \cos u \cdot \sin v$$
$$y = b \cdot \sin u \cdot \sin v$$
$$z = c \cdot \cos v$$

where $a = 1.2$, $b = 1.7$, $c = 2.1$, $u = 0, ..., 2\pi$, $v = 0, ..., \pi$.

    (b) Cross-cap

$$x = \cos u \times \sin 2v$$
$$y = \sin u \times \sin 2v$$
$$z = \cos^2 v - \cos^2 u \times \sin^2 v$$

where u and v range between 0 and π.

9. The turbulent-flow Nusselt number (Nu) for a plate of length l is a function of the Reynolds, Re, and the Prandtl, Pr, numbers:

$$Nu = \frac{0.037 Re^{0.8} Pr}{1 + 2.443 Re^{-0.3} \left(Pr^{2/3} - 1 \right)}$$

where $5 \times 10^5 \le Re \le 10^7$ and $0.6 \le Pr \le 2000$; use 100 values of Re and Nu each.

Plot $\log(Nu)$ as a function of $x = \log(Re)$ and $y = \log(Pr)$. Add axis labels and a caption.

10. The height of common prairie aster (in centimeters) plants was measured in the field, giving: 151, 174, 146, 155, 105, 122, 144, 154, 130, 151, 151, 154, 125, 140, 182, 160, 127, 141, 122, 140, 160, 157, 126, 129, 138, 156, 141, 182, 131, 156, 180, 180, 133, 162, 141, 189, 147. Plot a histogram for these data and add a title to the graph. Calculate the mean and standard deviation and write the determined values into the graph (with the `text` commands; take the initial text coordinates as follows: $x = 150$ and $y = 7$ for the mean, and $x = 150$ and $y = 6.5$ for the standard deviation).

11. Weight–height data for students in an academic group are (see Subsection 2.2.5.4): $h = 155, 175, 173, 175, 173, 162, 173, 188, 190, 173, 173, 185, 178, 168, 162, 185, 170, 180, 175, 180, 175, 175, 180, 165$ cm; $w = 54, 66, 66, 71, 68, 53, 61, 86, 92, 57, 59, 80, 70, 59, 50, 145, 68, 78, 67, 90, 75, 74, 84, 53$ kg. The data were fitted by the polynomial expression $w_f = 906.14 - 11.39h + 0.03780h^2$. Plot the $w_f(h)$ curve calculated by the fit expression and the data points. Take the heights for the fit expression in the interval between the minimal and maximal h-data values. Add axis labels, grid, a caption and a legend to the graph.

12. The M_2 molar concentration of a solute in water is calculated by the expression $M_2 = M_1 V_1 / V_2$ (see Subsection 2.2.5.4), where V_2 is the final solution volume, and M_1 and V_1 are the initial solute concentration and solution volume, respectively. Given the initial solution volume $V_1 = 0.1$ L, plot the 3D graph $M_2(M_1, V_2)$ in which M_1 changes in the range 0.5–2 mol/L and V_2 in the range 0.1–0.9 L. Show the 3D graph with azimuth and elevation angles of −135° and 35°, respectively. Add axis labels, a grid and a caption to the graph.

13. The partial pressure, P_A, of the gaseous A-component in the reaction A→B+C can be calculated by the expression $P_A = P_0 e^{-kt}$ where the initial partial pressure $P_0 = 55$ Torr, the reaction rate constant $k = 3$ min^{-1} and the time t changes from 1 to 5 min in steps of 0.25 min. Plot

two graphs on one page: $\ln(P_A)$ as a function of t, and $P_A(t)$. Add axis labels, a grid and a caption to each graph.

14. The molecular velocity distribution of gaseous helium is a function of the velocity v and temperature T and can be described by the Maxwell–Boltzmann equation:

$$f_v = 4\pi \left(\frac{M}{2\pi RT}\right)^{\frac{3}{2}} v^2 e^{-\frac{Mv^2}{2RT}}$$

where the molecular weight M is 0.004 kg/mol, the gas constant $R = 8.314$ J/(mol K), and v and T are in the ranges 0–1300 m/s and 50–500 K, respectively. Plot the graph $f_v(v,T)$ with axis labels, a grid and a caption.

15. In bio-fluid mechanics the motion of a rigid particle is described by time-dependent coordinate equations. Plot the trajectory of a particle with coordinates:

$$x = \cos(t)$$
$$y = \sin(t)$$
$$z = t^2$$

Use the `LineWidth` property equal to 5. The time t is given in 0–2 dimensionless units. Add axis labels, a grid and a caption.

16. The relationship between the surface coverage θ of an adsorbed gaseous substance on the substance pressure P above the surface at constant temperature (Langmuir isotherm) is:

$$\theta = \frac{bP}{1 + bP}$$

where b is constant for the given substance at given temperature. Plot the graph of $\theta(P,b)$ when P and b are in the range 0–1 and 0.1–100, respectively. Add graph axis labels, a grid and a caption.

17. The surface roughness height h of bio-films used in ophthalmology (lenses) and other medical applications can be modeled by the equation

$$h = a\sin(\sin 2\pi k(x + y))$$

where x and y are dimensionless surface size and change in the range 0–1 at steps of 0.01; k is the wave number and is equal to 5.

Plot a rough surface graph $h(x,y)$ with azimuth and elevation angles of –24° and 82°, respectively; add axis labels, grid and a caption to the graph.

3.6 Answers to selected exercises

3. (b) placing several plots on the same page.

5.

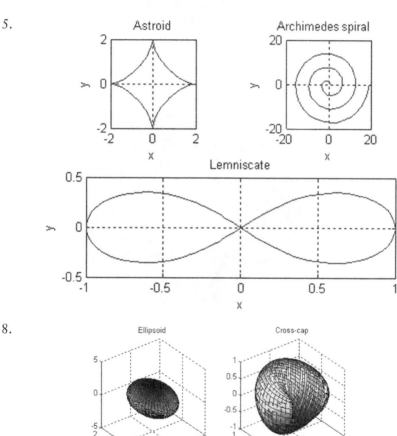

8.

12.

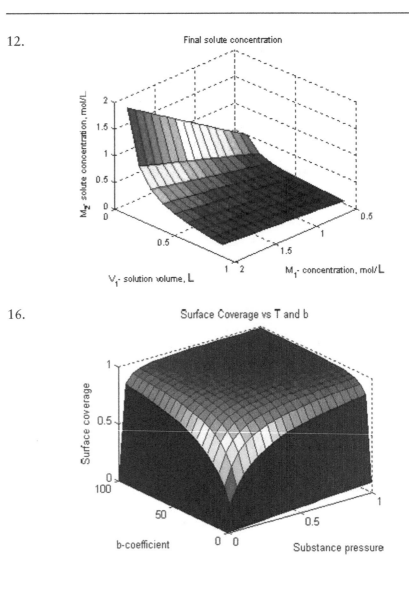

16.

4

Script, function files and some useful MATLAB® functions

The commands discussed in previous chapters were executed in the Command Window; they were not saved, and cannot be re-used. Another disadvantage of the Command Window is that correction is possible only on the last line executed.

When one of the sequence commands needs to be corrected, all its predecessors have to be re-executed and the command then corrected, after which all subsequent commands must be repeated to obtain the final result. Moreover, every time a calculation needs to be repeated, all commands have to be retyped and re-entered. This is inconvenient, and the reader who has perused the preceding chapters has undoubtedly experienced it. The remedy lies in writing all commands sequentially into a file, saving it and running it as necessary. MATLAB® has two types of files: script and function files. These types are explained further below. In addition, some MATLAB® functions of numerical analysis that are frequently used in applied calculations are presented.

4.1 Script files

4.1.1 Creation, saving and running

A script file is a sequence of commands (often called a program) that should be run. MATLAB® executes commands in the order they were written. Corrections can be entered directly in the file and new commands can be added. The extension of the saved script file is '.m', hence the term m-files.

A script file should be typed and edited into a MATLAB® Editor. To open the Editor, the command edit should be typed and entered in the

Command Window, or the 'New' and 'M-file' options in the 'File' menu should be selected. After this the Editor Window appears (Figure 4.1):

The commands of the script file should be typed line by line. It is also possible to type two or more commands on the same line, dividing them by commas or by semicolons. A new line will be accessible after pressing the Enter key. Each new line is numbered automatically. A command can also be typed in another editor and then copied into the Editor Window.

Figure 4.2 on p. 97 shows an elementary script file typed into the Editor Window. The first lines are usually explanatory comments, preceded by the comments sign % and in green (light gray here); they are not part of the execution. The commands appear in black for better legibility.

In the latest MATLAB® version the Editor Window appears with a message bar (located on the right) of the so-called M-Lint analyzer, which detects possible errors, comments on them and recommends modifications for better program performance. The bar is topped by the message indicator ▣. A green indicator means no errors, warnings or possibility of improvements; a red indicator means syntax errors have been detected; and an orange indicator means warnings or possibility of improvements (but no errors). When the analyzer detects an error or possibility of improvement, it underlines/highlights the text and a horizontal line appears on the message bar. When the cursor is placed at this line the comment message appears (see Figure 4.2). The changes it recommends – such as the one shown in this

Figure 4.1   The Editor Window.

Published by Woodhead Publishing Limited

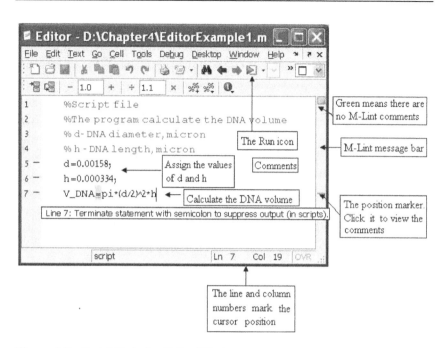

Figure 4.2 Script file in the Editor Window.

figure (addition of a semicolon) – need not be executed because we want to display the resulting value of the DNA volume, V_DNA. The M-Lint regime appears in the Editor Window by default and can be disabled by un-signing the 'Enable integrated M-Lint warning and error messages' check-box in the M-Lint option of the Preferences Window, which can be selected from the File in the Editor Window or in the Desktop menu.

The script file should be saved after it has been written into the Editor Window. For this the 'Save As ...' option from the File menu should be chosen. After this the 'Save file as ...' window appears and the desired file location and file name should be typed respectively into the 'Save in:' and 'File name:' positions. By default the 'MATLAB' subdirectory in the 'My documents' directory is automatically selected for the file location, or the desired directory has to be chosen from the pop-up menu which appears after pressing on the arrow on the right-hand side of the 'Save in:' field of the 'Save file as ...' window. The name for a script file should begin with a letter, and should not repeat the user-defined or predefined variables, MATLAB® commands or functions. In addition, it is not recommended to introduce spaces into the name, which cannot be longer than 63 characters.

To execute a script file: first, the user should check the current MATLAB directory, and if the file is not in it a directory with this file should be set; secondly, the file name should be typed in the Command Window and then entered.

Current directory

The currently set directory is shown in the 'Current Folder' field of the MATLAB® Desktop (see Figure 4.3). The simplest way to achieve the desired setting is:

- click on the icon with three points ⬚ on the right-hand side of the Current Folder field, after which the Browse for Folder Window appears;
- select and click on the desired directory which appears in the Current Folder field.

For example, the script file with name EditorExample1.m (Figure 4.2) was saved in the Chapter 4 folder and located on disk D. To run it we need to change the current directory to D:\Chapter4 in the described way, type the file name (without the m-extension) in the Command Window, and then press the Enter key (p.117, top).

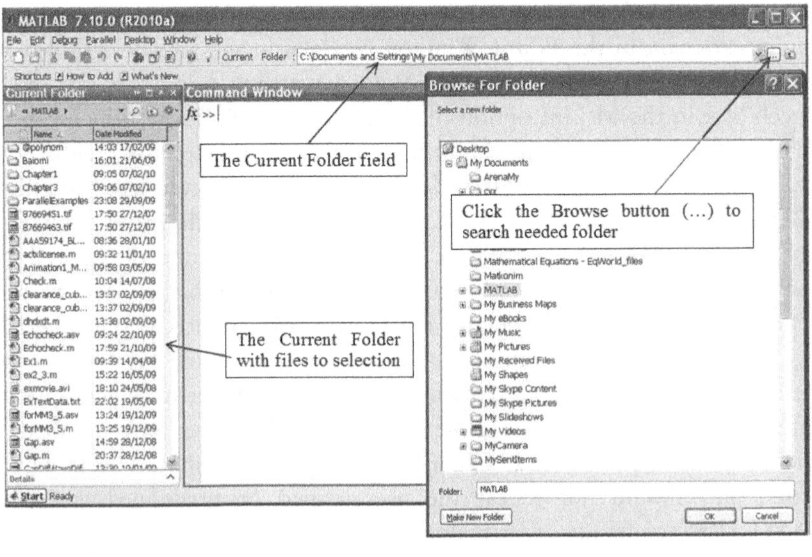

Figure 4.3 The 'Current Folder' field and 'Browse for Folder' Window.

Published by Woodhead Publishing Limited

>> runs the script file with the name EditorExample1
>> EditorExample1
 V_DNA =
6.5486e-010

The result of calculations via the commands is saved in the script file, and is displayed in the Command Window.

4.1.2 Input values to variables of a script file

To assign a value to the variables defined in the script file when the script is run the `input` command should be used; this command has the forms:

```
Numeric_Variable_Name = input('Displayed string')
Character_Variable_Name=input('Displayed string','s')
```

where `'Displayed string'` is a text that displays in the Command Window and prompts us to assign a number to the `Numeric_Variable_Name`, or a string to the `Character_Variable_Name`; `'s'` indicates that inputting characters are string.

After running a script file, when the `input` command is started, the string written in this command appears on the screen and the user should type and enter a number or string; inputted values are assigned to the `Numeric_Variable_Name` or to the `Character_Variable_Name` depending on the command form used.

As an example is a command script file that convert a weight, g_lb, in pounds to kilograms, g_kg, by the expression g_kg = g_lb/2.2046:

%Pound to kilogram convertor
g_lb=input('Enter your weight in pounds, G = ');
g_kg=g_lb/2.2046;
fprintf('\n Your weight is %5.1f kg\n',g_kg)

The file is saved with the `pound2kg` name. After typing and entering this file name in the Command Window, the prompt 'Enter your weight in pounds, G =' appears on the screen, a weight value should be typed (in pounds) after the sign '=' and `enter` pressed; the weight is converted to kilograms and displayed on the screen. The run command, prompts, and input and output numbers are:

>> pound2kg
Enter your weight in pounds, G = 189
 Your weight is 85.7 kg

Published by Woodhead Publishing Limited

With the input command the vectors and matrices can be inputted in the same way as for a variable – numbers in square brackets.

4.2 Functions and function files

4.2.1 Creating the function

In algebra a function is presented as the equation $y = f(x)$, whose right-hand side contains one – x – or more parameters (arguments) – x or x, a, b, ... When these parameters are assigned, the value of y is obtained. Similarly to the above, many MATLAB® commands discussed in the preceding chapters are written in function form, $\sin(x)$, $\cos(x)$, $\log(x)$, $\mathrm{sqrt}(x)$, etc., and can be used for direct calculations or placed into more complicated expressions by typing their name with the argument. In MATLAB® it is possible to create any new function and re-use it with different argument values and in different programs; such functions are termed 'user-defined'. Not only a specific expression, but a complete program created by the user can be defined as a function and saved as a function file. The MATLAB® definition of a function reads

```
function[output_parameters]=function_name(input_parameters)
```

The word `function` appears in blue and must be the first word in the file. The function name is written to the right of the '=' sign and should obey the same rules as for variable names (see subsection 2.1.4). As `input_parameters` we should write a list of arguments for transfer into the function, and as `output_parameters` a list of those we want to derive from the function. The input parameters must be written in parentheses and the output parameters in square brackets. (In the case of a single output parameter the brackets can be dispensed with.) Parameters must be divided by commas.

As with a script file, a function file should be written in the Editor Window. An example of a function file is presented in Figure 4.4 on p. 101. This function is named `dilution` and calculates the molar concentration of a solute after dilution. It has three input parameters – the molar concentration M_1 and the solution volume V_1 before dilution, and solution volume V_2 after dilution – and one output parameter – the molar concentration of the solute after dilution.

The requirements and recommendations regarding the structure and certain parts of the function file are:

Published by Woodhead Publishing Limited

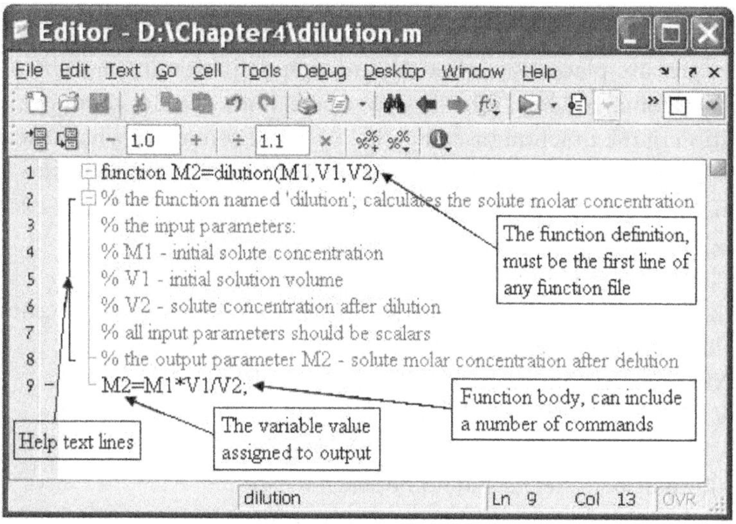

Figure 4.4 Typical function file in the Editor Window.

Line with the function definition

In addition to the previous comments regarding the function definition, the function can be written with the parameters omitted completely or in part. Possible variants of the function definition line are given in the following examples:

```
function [A, B]=example1(a, b, c)
```
full record, function name 'example1', three input and two output parameters;

```
function [A, B, C]=example2
```
function without input parameters, function name 'example2', three output parameters;

```
function example3(a, b, c)
```
function without output parameters, function name 'example3', three input parameters.

```
function example4
```
the function without parameters, function name 'example4'.

Number and names of the input and output parameters can differ from those in the examples. The word 'function' should be written in lower-case letters.

Lines with help comments

These lines are placed just after the function definition line and before the first non-comment line. The first of these lines should contain a short definition of the function (used by the `lookfor` command when it searches the information). The help comments are displayed when the `help` command is introduced with the user-defined function name; for example, typing into the Command Window `help dilution` yields:

>> help dilution

the function named 'dilution'; calculates the solute molar concentration

the input parameters:

M1 – initial solute concentration

V1 – initial solution volume

V2 – solute concentration after dilution

all input parameters should be scalars

the output parameter M2 – molar concentration after dilution

Lines of function body, local and global variables

The function body can contain one or more commands for actual calculations; between these commands, usually at their end, should be placed the assignments to the output parameters; for example, in the example in Figure 4.4 the output parameter is M_2 and the last command calculates and assigns the found value to M_2. The actual values must be assigned to the input parameters before running the function.

The variables in the function file are local and relevant only within the file. This means that once the function has been run, the variables are not saved and no longer appear in the workspace. If we want to share some or all of them with other function(s), we should make them accessible, which can be done with the `global` command in the form

```
global variable_name1 variable_name2 …
```

The global command must be written before the variable is first used in the function and also in other functions where it is intended to be used.

4.2.2   Saving a function as a function file

Before a function is used, it must be saved in a file. This is done exactly as for a script file – selecting 'Save As' from the File menu and then entering the desired location and name for the file. It is recommended that the file

Published by Woodhead Publishing Limited

be called by the function name, e.g. the `dilute` function should be saved in a file named `dilute.m`.

Examples of function definition lines and function file names:

`function a=fly(h,l)`	the function file should be named and saved as `fly.m`;
`function [v, m]=dna_r(d,h)`	the function file should be `dna_r.m`;
`function weights(h)`	the function file should be `weights.m`.

When a program has two or more functions, the name of a function file should be identical to the name of the main function (from which the program starts).

4.2.3 Running a function file

The function file can be run from another file or from the Command Window as follows. The function file definition line should be typed without the word 'function' and the input parameters should be assigned their values. For example, the `dilution` function file (see Figure 4.4) can be run thus:

```
>> M2=dilution(0.1,0.5,0.6)
M2 =
0.0833
```

Or alternatively with pre-assignment of the input variable:

```
>> a=0.1;b=0.5;c=0.6;
>> M2=dilution(a,b,c)
M2 =
0.0833
```

A user-defined function can be used in another mathematical expression or program; for example, `dilution` gives M_2 solute concentrations in mol/L for $V_2 = 0.6$ L of the solution; thus the solution has M2_mol =M2*V2 moles. For the calculations the following commands should be typed in the Command Window:

```
>>% defines values for the input parameters in the dilution function
>> a=0.1;b=0.5;c=0.6;
>> M2_mol=c*dilution(a,b,c)     % moles in the 0.6 L solution
M2_mol =
0.0500
```

Comparison of script and function files

First-year students, and life science students in particular, may find it difficult to understand the differences between script and function files, because most of their problems can be solved via the former type, or simply by inputting commands directly into the Command Window. Accordingly, the similarities and differences between these types are outlined below.

- Both are produced with the help of the Editor and saved as m-files (files with the extension m)
- The first line of the function file must be the function definition line
- The name of the function file should be the same as that of the function
- Function files can receive and return data through the input and output parameters, respectively
- Script files require parameters to be written straight into the file or with input commands
- Function files can be used as functions in other MATLAB® files or simply in the Command Window.

4.3   Some useful MATLAB® functions

Biotechnology laboratory practice involves a variety of mathematical operations such as interpolation, extrapolation, solution of algebraic equations, integration, differentiation and fitting. MATLAB® functions that can be used for these purposes are discussed below.

4.3.1   Interpolation and extrapolation

When data are available (by measurement or by tables) at certain points and we need to estimate intermediate values between them, this can be done by interpolation. When values outside the data points need to be evaluated, this can be done by extrapolation. For example, the values of measured radioactive decay R of a liquid substance are 400, 200, 100, 50 and 20 disintegrations per minute at times $t = 1, 2, 3, 4$ and 5 hours, respectively. The graph with the data points is presented in Figure 4.5. Finding the decay values at $t = 1.2$ and 2.4 hours (values between data points) is an

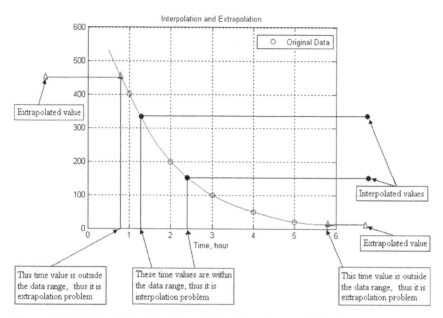

Figure 4.5 Original data (O), interpolation (●) and extrapolation (Δ) points.

interpolation problem, and finding them at $t = 0.8$ and 5.8 hours (values outside data points) is an extrapolation problem.

In MATLAB® both interpolation and extrapolation operations can be done with the `interp1` function. This command has the forms

```
yi=interp1(x,y,xi,'method') or
yi=interp1(x,y,xi,'method','extrap')
```

The first version is used for interpolation, and the second for extrapolation or for simultaneous interpolation and extrapolation.

The output parameter for these MATLAB® functions is `yi` – the interpolated value or vector of values. The input parameters are `x` and `y`, vectors of data values – argument and function, respectively; `xi`, a scalar or a vector with the points for which the values of `yi` are sought; `'method'` is the string that contains the name of the mathematical method to be used for interpolation or extrapolation; some available methods are `'linear'`, `'cubic'` and `'spline'`; a default method is `'linear'`, which need not be specified.

Published by Woodhead Publishing Limited

The 'extrap' string is required for the 'linear' method; the word can be omitted for the 'cubic' and 'spline' methods.

For the example in Figure 4.5 the commands for interpolation (at points 1.2 and 2.4 hours) and extrapolation (at points 0.8 and 5.8 hours) are:

```
>>R=[400, 200, 100, 50, 20];        % the data for x
>>t=1:5;                            % the data for y
>>xi=[1.2 2.4];                     % the points of interpolation
>>xe=[0.8 5.8];                     % the points of extrapolation
>>y_interpolated=interp1           % defines interpolated values
  (t,R,xi,'spline');
>>y_extrapolated=interp1           % defines extrapolated values
  (t,R,xi,'linear','extrap');
>> display interpolated and extrapolated values
>> y_interpolated,y_extrapolated
  y_interpolated =
  349.3600 151.0800
  y_extrapolated =
  360 160
```

4.3.2   Solution of non-linear algebraic equations

The matrix solution for a set of linear algebraic equations was discussed in chapter 2 (subsections 2.2.2, 2.3.4.1). Here we deal with non-linear equations with iterations in which the computer finds the x value at which $f(x) = 0$. The function for this process is fzero and its general form is:

```
x=fzero('fun', x0)
```

The fzero function seeks an x value representing the solution near the guess point x_0; and 'fun' is the string function to be solved. In this string the equation itself can be written, or the name of the user-defined function that contains it.

One possible way to determine the approximate value of x_0 is to plot the graph of the equation and check the x-value at which the function is zero.

For example, consider the Arrhenius equation

$$\frac{k}{A} = e^{\frac{-E_0}{RT}}$$

where the value $k/A = 8.7271 \times 10^{-14}$ was previously determined at $T = 300$ K and $R = 8.314$ J/kmol. The unknown variable is the activation energy E_0 and should be named x, and the equation presented in the form

$f(x) = \dfrac{k}{A} - e^{-\frac{x}{RT}}$. The value of x_0 must be positive for physical reasons and for many reactions it is around 5×10^4 J/mol.

The solution command is:

```
>> Eo=fzero('8.7271e-14-exp(-x/(8.314*300))',1e4)
Eo =
7.5000e+004
```

The string with the solved equation cannot include pre-assigned variables; for example, it is not possible to define $k/A = 8.7271 \times 10^{-14}$, $R = 8.314$, $T = 300$ K and then write 'k_A-exp(-x/(R*T))'.

For substitution of pre-assigned parameters into the solving expression the following form of fzero is available:

```
x=fzero(@(x) fun(x, preassigned_variables1,
        preassigned_variables2,… ), x0)
```

$@(x)$ fun[1] indicates a user-defined function that contains the equation with the additional arguments preassigned_variables1, preassigned_variables2, … .

Using this form, we can write the above example as a function file with the name actener, input parameters k/A, R, T and x_0, and output parameter E_0. In this function file the fzero command reads x=fzero (@ (x) myf(x,k_A,R,T), x0) and the function definition line for the myf function should appear as function f=myf(x,k_A,R,T). The full text of the actener function is

```
function Eo=actener(k_A, R, T,x0)
% activation energy calculation
% k_A is reaction constant
% R – gas constant, J/mol/K
% T – temperature, K
% x0 – initial Eo
Eo=fzero(@ (x) myf(x,k_A,R,T), x0);
function f=myf(x,k_A, R, T)
f=k_A-exp(-x/(R*T));
```

Published by Woodhead Publishing Limited

This function should be saved as `actener.m` file and run into the Command Window with assigned parameters k/A, R, T and x_0:

```
>> Eo=actener(8.7271e-14, 8.314, 300,5e4)
Eo =
7.5000e+004
```

4.3.3 Integration

The area beneath a segment of a function $f(x)$ or beneath data points (Figure 4.6) can be determined by numeric integration. For this the area is subdivided into small geometrical elements, e.g. rectangles and trapezoids. The integral is the sum of the areas of these elements. Below, two MATLAB® functions for integration are described – quad and trapz. The first is used when $f(x)$ is presented as an analytical expression, and the second when it is presented as data points.

The quad function

This function uses the adaptive Simpson method and has the form:

```
q=quad('function',a,b,tol)
```

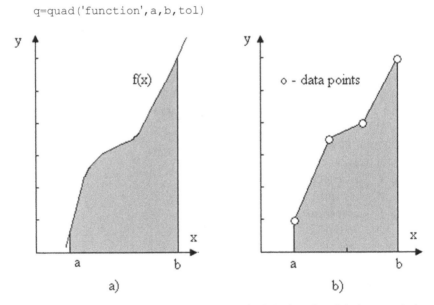

Figure 4.6 Definite integral (the shaded area) of the function $f(x)$ given analytically (a) and by the data points (b).

Published by Woodhead Publishing Limited

where 'function' is a string with the expression $f(x)$ to be integrated, or the name of a function file where the expression should be written (in the same way as for the fzero command); a and b are the integration limits, tol is the desired maximal absolute error (this parameter is optional and can be dispensed with); the default tolerance is smaller than 1×10^{-6}; q is the variable to which the value of the integral obtained is assigned.

The function $f(x)$ should be written using element-wise operators, e.g. '. *', '. ^' and '. /'.

For example, to calculate the integral $q = \int_1^5 \frac{1}{x^2} dx$ (the function $1/x^2$ appears in second-order reaction rates, in population dynamics, etc.) the following command should be typed and entered from the Command Window

```
>> q=quad('1./x.^2',1,5)     % Note: the element-wise operations are
                               used for 1/x²

q =
0.8000
```

Another possible way of using quad is to write in the Editor Window and save the function file that contains the function $1/x^2$:

```
function y= Integr_Ex(x)
y=1./x.^2;
```

To integrate the saved function in the Integr_Ex.m file, the quad command should be typed and then entered from the Command Window:

```
>> q=quad('Integr_Ex',1,5)
q =
0.8000
```

Preassigned parameters can be substituted in the 'function' expression in the same way as for the fzero command.

The trapz function

The form of this function is

```
q=trapz(x,y)
```

where x and y are vectors of the point coordinates; the function uses the trapezoid method for numerical integration when the function is presented as data points.

Published by Woodhead Publishing Limited

For example, for growth rates of:

t (h)	0	1	2	3	4	5	6
v (cell/h)	1	2	5	11	22	44	178

the number N of bacterial cells at the end of the measurements is

$$N = N_0 + \int_0^6 v(t)\,dt$$

where $N_0 = 10$ is the initial cell number; $v(t)$ under the integral is given numerically in the table. The result, which obviously cannot be a fraction, should be rounded to the nearest lesser integer.

The next command should be typed and entered from the Command Window:

```
>>N0=10;
>>t=0:6;
>>v=[1 2 5 11 22 44 178];
>>% the floor rounds number towards minus infinity
>>N=N0+floor(trapz(t,v))
N =
183
```

4.3.4 Derivatives

The derivative dy/dt of the function $y(t)$ is characterized by the infinitesimal change in the function at some given point; in geometrical representation (Figure 4.7, p. 111) it is the slope of the tangent to the curve at the ith point (x,t).

Following these definitions, the derivative can be presented for numerical calculations as:

$$\frac{dy}{dt} = \lim_{t \to \infty} \frac{\Delta y}{\Delta t} \approx \frac{y_{i+1} - y_i}{t_{i+1} - t_i}$$

Thus, the derivative can be calculated at each i-point as the ratio of y- and x-differences between neighboring points. The diff command for these differences has the forms:

```
dy=diff(y) or dy_n=diff(y,n)
```

where y is the vector of y-values at points $i = 1, 2, \ldots$; the n argument indicates how many times the diff command should be applied and the

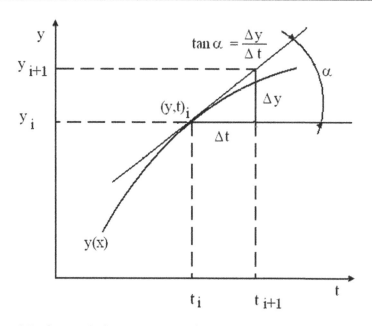

Figure 4.7   Geometrical representation of the derivative.

order of the particular difference. For example, if n = 2 the diff command should be applied twice; the dy and dy_n output parameters are vectors of the determined values of the first- and nth-order differences.

The dy vector is one element shorter than y and correspondingly the dy_n vector is *n* elements shorter than y; for example, if *y* has 10 elements dy has nine elements and, say, the third-order derivative vector dy_3 has only seven elements.

For example, a species concentration *c* changes with time *t* due to a chemical reaction, as shown here:

t (min)	0	1	2	3	4	5	6	7
c (mol/L)	0.2500	0.166	0.1250	0.1000	0.0833	0.0714	0.0625	0.0556

These data have to be used to determine the reaction rate $r = -dc/dt$ [mol/(L min)]. The commands to be applied in the Command Window are:

```
>> t=0:7;
>> c=[0.2500 0.1667 0.1250 0.1000 0.0833 0.0714 0.0625 0.0556];
```

```
>>dt=diff(t);dc=diff(c); %t and c differences
>> r=-diff(c)./diff(t);% element-wise division
>> disp('t          r          '),disp([t(2:end)' r'])
          t               r
     1.0000       0.0833
     2.0000       0.0417
     3.0000       0.0250
     4.0000       0.0167
     5.0000       0.0119
     6.0000       0.0089
     7.0000       0.0069
```

The second-order derivative, representing the acceleration rate of the reaction, $a = \dfrac{dr}{dt} = -\dfrac{d^2c}{dt^2} = -\dfrac{\Delta c_i - \Delta c_{i-1}}{\Delta t^2}$.

is:

```
>>a=-diff(r)./diff(t(2:end)) % t(2:end) is used as r is one element shorter
a =
-0.0416 -0.0167 -0.0083 -0.0048 -0.0030 -0.0020
```

Negative values of a indicate deceleration of the reaction. The same results can be reached via the second-order c differences:

```
>>a=-diff(c,2)./diff(t(2:end)).^2
a =
-0.0416 -0.0167 -0.0083 -0.0048 -0.0030 -0.0020
```

Note, that if the step h of argument t is constant, it can be used in changing diff(t).

The derivatives of a function given by an equation can be determined in the same way as tabulated data; the step in t in such a case can be smaller and lead to more exact derivative values. For example, assuming that the data in the above table are very accurately described by the equation c = 0.25/ (1 + 05t), the reaction rate can be determined with the following commands:

```
>> h=0.5;
>>t=0:h:7;c=0.25./(1+0.5*t);
>> r=-diff(c)./h;
>> disp('   t        r'),disp([t(2:end)' r'])
```

t	r
0.5000	0.1000
1.0000	0.0667
1.5000	0.0476
2.0000	0.0357
2.5000	0.0278
3.0000	0.0222
3.5000	0.0182
4.0000	0.0152
4.5000	0.0128
5.0000	0.0110
5.5000	0.0095
6.0000	0.0083
6.5000	0.0074
7.0000	0.0065

Significant deviations in the values of r and t are evidence that the number of points in the table is not enough for exact determination of the derivatives. They also confirm that for an analytically given function, more points apparently provide more exact results.

Frequently, a problem involving a derivative is formulated so that the latter (reaction rate, population growth, etc.) needs to be determined for one specific point. In such a case the minimal range of t (or any other possible argument $-x$, z, etc.) should be assigned one step after that point. For example, if the required value of a rate is needed at $t = 1$ hour, the minimal interval for the numerical differentiation is $t \dots t + h$, where h is the time step, which should be specified.

4.3.5 Curve fitting

The process of matching an expression or curve to the data points is called curve fitting or regression analysis. The fitting expression may be based on a theory or chosen empirically. This subsection describes MATLAB® curve fitting involving polynomials.

The process consists of finding the values of the coefficients of the mathematical expression which should fit the experimental data. In the case of a polynomial expression

$$y = a_{n-1}x^{n-1} + a_{n-2}x^{n-2} + \dots + a_1 x + a_0$$

the coefficients are the a_i values, and n here is both the length of the polynomial and the number of coefficients in it.

The highest exponent, $n - 1$, is called the degree of the polynomial; for example, the strait line $y = a_1x + a_0$ is a first-degree polynomial that has two coefficients, a_1 and a_0, to be determined.

When the experimental data contain n points (x,y) the n coefficients can be determined by solving the set of $y(x)$ equations written for each of the points. Polynomial fitting where the number of coefficients is the same as that of points is not always efficacious and might lead to significant discrepancies in regions between data points. Another case of possible inefficiency is a linear $y(x)$ dependence where two points suffice for determination but the experimental data contain a larger number, which means more equations than unknowns. The commonly used method for better fit is least squares, in which the coefficients are determined by minimizing the sum of squares of the differences between the polynomial and data values; these differences are called residuals and are denoted R. The minimization is done by taking the partial derivative of R with respect to each coefficient and imposing their equality to zero, for example for the straight line $\frac{\partial \sum R_i^2}{\partial a_1} = 0$ and $\frac{\partial \sum R_i^2}{\partial a_0} = 0$, and the set of two equations should be solved to determine a_1 and a_2.

Polynomial fitting in MATLAB® can be done with the `polyfit` function, whose simplest form is

```
a=polyfit(x,y,m)
```

where the input arguments x and y are the vectors of the x- and y-coordinates of the data points and m is the degree of the polynomial, which in the notation used above is equal to $n - 1$; the output argument a is the $(m + 1)$-element vector of the fitting coefficients, in which the vector of the first element a(1) is a_{n-1}, the second a(2) is a_{n-2}, ..., and the final a(m+1) is a_0.

With the fitting coefficients the y-values can be calculated at any x in the fitted interval with the `polyval` command; this command has the following simplest form

```
y_polynomial=polyval(a,x_polynomial)
```

where a is the vector of polynomial coefficients as per the `polyfit`, and x_polynomial is that of the coordinates at which the y_polynomial values are calculated.

Published by Woodhead Publishing Limited

For example, the height of five seedlings and the amount of nutrients for each of them are 4.7, 7, 15.2, 22.3 and 27.2 cm and 12, 28, 60, 95, 120 mg. For fitting these data by the first-degree polynomial ($m = 1$), and for generating the next curve and data plot, the following script file, named fitexample, can be created:

```
%Curve fitting
% x- seedling height, cm
% h - nutrient amount, mg
x=[12,28,60,95,120];        % vector of x with nutrient data
y=[4.7,7,15.2,22.3,27.2];   % vector of y with seedling data
a=polyfit(x,y,1)            % defines fitting coefficients
x_pol=12:0.5:120;           % vector of x values for plotting
y_pol =polyval(a,x_pol);    % vector of y values calculated at x_pol
plot(x,y, 'o', x_pol, y_pol)  % plots polynomial and data points
xlabel('Nutrient, mg'), ylabel('Seedling height, cm'),grid
legend('original data', 'first degree polynomial fit', 'location', 'best')
```

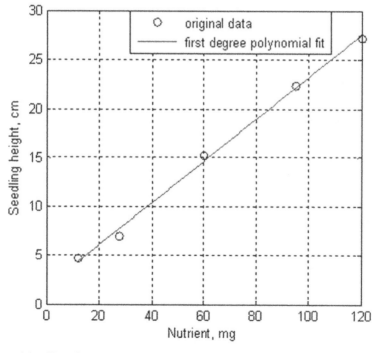

Figure 4.8 First-degree polynomial fit.

After typing and entering the file name in the Command Window, the following coefficients are displayed and the graph in Figure 4.8 is generated:

>> fitexample

a =

0.2138   1.8090

Many bio-applications require an exponential or a logarithmic fitting function:

$$y = a_0 e^{a_1 x}$$
$$y = a_1 \ln x + a_0 \quad \text{or} \quad a_1 \log x + a_0$$

In the first case the fitting function should be rewritten as $\ln y = a_1 x + \ln a_0$ and the `polyfit` function can be used with argument $\log(y)$ instead of y, e.g. a=polyfit(x, log(y),1) where in the vector a the a(1) element is a_1 and a(2) is $\ln a_0$, thus $a_0 = \exp(a(2))$.

In the second case the `polyfit` function can be used with argument $\log(x)$ or $\log10(x)$ instead of x, e.g. a=polyfit(log(x),y,1) or a=polyfit(log10(x),y,1) where in vector a the a(1) element is a_1 and a(2) is a_0.

4.4 Application examples

4.4.1 Body mass index

Body mass index (BMI) I is a measure that relates the body weight G in kg and height h in m; it is defined as

$$I = G/h^2.$$

This index is frequently used to assess how much an individual's body weight departs from what is normal or desirable. A BMI between 20 and 25 kg/m² is normal, below 20 underweight, between 25 and 30 overweight, and above 30 obese.

Problem: Compose a script program that calculates and displays the BMI of the user and writes a conclusion about the user's weight.

In the script presented below, the user enters his/her weight and height, and the program then calculates the BMI, determines the user's category and displays the results.

Published by Woodhead Publishing Limited

```
% BMI calculator
clear
w=input('Enter your weight in kg');
h= input('Enter your height in m');
I=w/h^2;
if I<20
 S='too small';
elseif I>=20&I<=25
 S='normal';
elseif I>=25&I<=30
 S='large';
else
 S='too large';
end
fprintf('\n Your BMI is %5.1f kg/m^2, thus your weight is %s\n',I,S)
```

With the script file name BMI entered into the Command Window, the program displays the prompt to input the weight in kg; with the weight typed and the Enter key pressed, a new prompt for height in m appears; with the latter typed and entered, the BMI is calculated and displayed together with a conclusion.

The following shows the Command Window when the file was run:
```
>> BMI
Enter your weight in kg 85
Enter your height in m 1.8
Your BMI is 26.2 kg/m^2, thus your weight is large
```

4.4.2 Basal metabolic rate

The basal metabolic rate (BMR) P in kcal/day is the human minimal daily energy requirement (in calories). For women this index is given by the expression:

$$P = 655 + 9.6w + 1.8h - 4.7a$$

and for men by:

$$P = 66 + 13.7w + 5h - 6.8a$$

where w is weight in kg, h is height in cm and a is age in years.

Published by Woodhead Publishing Limited

Problem: Write a script program for calculating and displaying the BMR. The script file solving this problem is written below.

```
% BMR calculator
clear
sex=input('Enter the sex: 1 - male, 2 - female');
w=input('Enter your weight in kg');
h= input('Enter your height in m ');
a=input('Enter your age in years ');
if sex<1
P=66+13.7*w+58*h-6.8*a;
else
P=655+9.6*w+1.8*h-4.7*a;
end
fprintf('\n Your BMR is %5.1f cal/day\n',P)
```

After entering into the Command Window the script file name BMR, the program displays the prompt to sex, type 1 or 2 (male or female). After pressing the enter key, prompts then appear successively about weight in kg, height in m and finally age. The BMR is then calculated and displayed. The following shows the Command Window when the file was run:

```
>> BMR
Enter the sex: 1 - male, 2 - female 2
Enter your weight in kg 85
Enter your height in m 1.72
Enter your age in years 64

Your BMR is 1173.3 cal/day
```

4.4.3 Conversion of concentration units

The concentration of components in solutions is expressed in different forms, among them molarity, molality and molar fraction. Molarity M is the solute amount per liter of solution, mol/L; molality m is the solute amount per solvent mass, mol/kg; and molar fraction χ is the mole solute to mole solution ratio. The expression for conversion from M to m and to χ is

$$m = \frac{1000M}{1000d - MM_1}$$

$$\chi = \frac{m}{m + 1000M_2}$$

where d is the solution density (g/L), M_1 the molecular weight of the solute (g/mol) and $M_2 = 18$ g/mol is the molecular weight of water.

<u>Problem:</u> Write a function file with name `molfrac` that calculates and displays solute molality and molar fraction for a given solute molarity and molar mass, and solution density. Calculate the molality and the molar fraction for these data: $M = 3$ mol/L, $d = 1.5 \times 10^3$ g/L and $M_1 = 40$ g/mol (sodium hydroxide).

The input parameters in the function definition line are: solution density `d`, solute molarity `M` and molar mass of the solute `M1`; the output parameters are `molality` and `molar_fraction`. The function file is:

```
function [molality,molar_fraction]=molfrac(d,M,M1)
% molality (mol/l) and molar fraction calculations
% d is solution density in g/l
% M1 is solute molar mass in g/mol
M2=18; % g/mol, water
molality=1000*M/(1000*d-M*M1);
molar_fraction=molality/(molality+1000*M2);
```

The following shows the Command window when the file was run:

```
>> [molality,molar_fraction]=molfrac(1.5e3,3,40)
molality =
0.0020
molar_fraction =
1.1112e-007
```

4.4.4 Population of Europe

Population statistics for Europe are:

Year	1750	1800	1850	1900	1950	1999	2008
Population (millions)	163	203	276	408	547	729	732

<u>Problem:</u> Write a function file with name `europes` that predicts and displays the population in 2011. Use for prediction the `interp1` function with three extrapolation alternatives provided: 1, linear; 2, cubic; 3, spline.

The first line of the function should be the function definition line with the function name `europes` and the input and output parameters. Thus, the following input parameters are specified: `yi`, the target year, and `num`, the chosen method alternative: 1, `linear`; 2, `cubic`; 3, any other number (except 1 and 2) for the `spline`. The specified output parameter is `pei` – the predicted population.

The function file written to solve the problem is:

```
function pei=europes(yi,num)
% Europe population prediction
% yi - year for prediction
% num - mehod number: 1-linear,2 - cubic, 3 - spline
y=[1750:50:1950 1999 2008];
p=[163 203 276 408 547 729 732];
if num==1
 method='linear';
elseif num==2
 method='cubic';
else
 method='spline';
end
pei=interp1(y,p,yi,method, 'extrap');
```

The following appears in the Command Window when the file was run:

```
>> pei=europes(2011,1)
 pei =
 733
```

4.4.5 Reactant breakdown

In the two-stage consecutive reaction A→B→C, the concentration of intermediate reactant B can be expressed as

$$[B] = \frac{[A]_0\, k_1}{k_2 - k_1}\left(e^{-k_1 t} - e^{-k_2 t}\right)$$

where $[A]_0$ is the initial concentration of reactant A, k_1 and k_2 are the reaction rate constants for the A→B and B→C reactions, respectively, and t is time.

<u>Problem</u>: Write a function file `reactime.m` that evaluates and displays time t in the course of which the intermediate reactant B appears, accumulates and breaks down to the current concentration. Calculate the time for: $[B] = 0.01$ mol/L at initial concentration $[A]_0 = 2.5$ mol/L, and constants $k_1 = 2.6 \times 10^{-3}$ s^{-1} and $k_2 = 9.5 \times 10^{-3}$ s^{-1}; use for solution the `fzero` function with initial (guess) value $x_0 = 2$ s.

The input parameters for the function definition line are: the concentrations $[B]$ and $[A]_0$, the reaction rate constants k_1 and k_2, and the initial value x_0;

the output parameter is time t. In using `fzero`, the variable t in the expression above should be named x. The function file for solving this problem is:

```
function t=reactime(A0,B,k1,k2,x0)
% time of reactant breakdown, t
% Ao is initial concentration of the reactant A, mol/l
% B is the intermediate component concentration, mol/l
% k1 is reaction rate constant of the A-to-B reaction
% k2 is reaction rate constant of the B-to-C reaction
t=fzero(@(x) reaction(x,A0,B,k1,k2),x0);
function f=reaction(x,A0,B,k1,k2)
f=-B+A0*k1/(k2-k1)*(exp(-k1*x)-exp(-k2*x));
```

The following appears in the Command Window when the file was run:

```
>> t=reactime(2.5,0.01,2.6e-3,9.5e-3,2)
t =
1.5530
```

4.4.6 Amount of compound

The decay rate v of a compound is given by $v(t) = -0.69*e^{-2.3t}$, where t is time in h, and v is in mol/h. The amount m of the compound after time t is:

$$m = m_0 + \int_0^t v(t)\,dt$$

where $m_0 = 0.3$ mol is the initial mass of the compound.

Problem: Write a script that calculates the amount of the compound every 30 minutes over 3 hours and plot the graph $m(t)$.
   The solution is organized in the following steps:

- define the variables m_0 and t;
- organize a loop with `for ... end` command for the amount of compound, m, calculated at different upper limits; for integration the `quad` function is used, in which the upper integration limit repeatedly changes from 0.5 up to 3 at steps of 0.5 hours;
- the results are displayed with the `fprinntf` command;
- the graph is plotted with the `plot` command.

The script to solve this problem is:

```
% Amount of compound
clear, close all        % clear the memory and close all preceding
                        % graphs
```

```
m0=0.3                          % initial amount of compound
t=0.5:0.5:3;                    % times for upper integration limit
for i=1:length(t)               % the loop for integration by the quard
                                % command
  q(i)=m0+quad('-0.69*exp(-2.3*t)',0,t(i));
end
fprintf('\nCompound amount change\n Hour Mole\n')
                                % title for resulting table
fprintf('%5.1f %8.5f\n',[t;q])   % displays resulting table
plot(t,q, '-o')                 % plot the graph
title('Compound decay')
xlabel('Time, h'),ylabel('Mass,mol')
grid
```

This script was named and saved into the Integr2_Ex.m file. The following appears in the Command Window when the file was run:

```
>> Integr2_Ex
  Compound amount change
  Hour       Mole
  0.5        0.09499
  1.0        0.03008
  1.5        0.00952
  2.0        0.00302
  2.5        0.00095
  3.0        0.00030
```

And the resulting graph is:

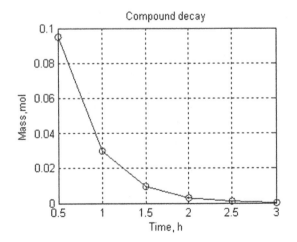

4.4.7 Dissociation rate

Species dissociation is described by the equation

$$m = m_0 e^{-kt}$$

where m_0 and m are the amounts of the species initially and at time t, respectively, and k is the dissociation constant.

<u>Problem</u>: Write and save a function that calculates the reaction rate $r = -dm/dt$ at $t = 0, 0.1, ..., 10$ min, $k = 1.8$ min^{-1} and $m_0 = 5$ mg, and plots the $m(t)$ and $r(t)$ curves in one graph.

The program is realized in the form of a function named `DissRate` with the following input arguments: `m0`, the species amount at starting time; `k`, the reaction constant; `tstart`, the starting time; `tend`, the end time; and `h`, the time step.

The solution is carried out in the following steps:

- the species amount vector `m` is calculated for the whole time interval – from `tstart` to `tend` with step `h`;
- the reaction rates are calculated as the `-diff(m)` to `h` ratio and assigned to the vector `r`;
- the resulting matrix is constructed, in which the first column is `t` and the second is `r`; the columns are adjusted to the same length, as `r` is one element shorter than `t`;
- the results are printed with the `disp` command;
- the $m(t)$ function and its derivative are generated with the `plot` command and then the commands for axis and plot captions are added; the vector r is one element shorter than the t and m vectors, and thus the `axis` command is introduced to show both $m(t)$ and dm/dt curves in the same t interval.

The function file to solve this problem is:

```
function DissRate(m0,k,tstart,tend,h)
% Species dissociation rate
% m0 is initial species amount, mg
% k is dissociation constant, 1/min
% tstart and tend are started and final values of time, min
% h is time step, min
t=tstart:h:tend;
m=m0*exp(-k*t);
```

```
r=-diff(m)./h;
t_r=[t(2:end)' r'];
disp('    t       r')
disp(t_r)
plot(t,m,t_r(:,1),t_r(:,2),'-- ')
xlabel('time'),ylabel('m and dm/dt'),grid
legend('Function m(t)','Dissociation Rate dm/dt')
title('Species Dissociation and Dissociation Rate')
axis([tstart tend-h 0 max(r)])
```

After running this file, the following appear in the Command and Figure windows:

```
>> DissRate(5,1.8,0,2,0.1)
     t          r
  0.1000     8.2365
  0.2000     6.8797
  0.3000     5.7464
  0.4000     4.7998
  0.5000     4.0091
  0.6000     3.3487
  0.7000     2.7971
  0.8000     2.3363
  0.9000     1.9515
  1.0000     1.6300
  1.1000     1.3615
  1.2000     1.1372
  1.3000     0.9499
  1.4000     0.7934
  1.5000     0.6627
  1.6000     0.5535
  1.7000     0.4624
  1.8000     0.3862
  1.9000     0.3226
  2.0000     0.2694
```

The data are plotted in the figure overleaf:

4.4.8 Fitting of cancer incidence data

A high level of a particular substance in the atmosphere may increase the incidence of cancer. Assume that the function that relates cancer incidence Y to the amount of carcinogen X is polynomial. Observed cancer incidences

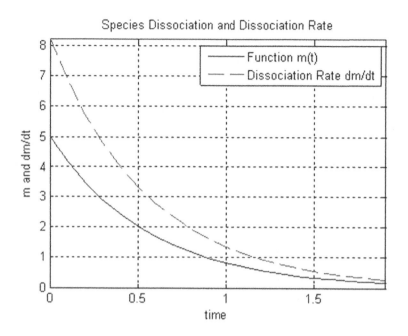

Species Dissociation and Dissociation Rate

are 10, 10, 30, 35, 55 and 120 at carcinogen concentrations respectively of 10, 40, 100, 200, 300 and 400 (dimensionless).

<u>Problem</u>: Write and save a function that requires degree of polynomial as input parameter, and then fits the cancer incidence–carcinogen concentration data, displays fitting coefficients and plots $y(x)$ points with defined polynomial curve. Take the degree of polynomial to be no more than 5.
The solution is:

```
function cancer(m)
%Cancer-concentration fit
%m - degree of polynom
%a - fitting coefficients
if m>5
 disp('Error: Polynomial degree can not be more than 5')
else
 x=[10 40 100 200 300 400];
 y=[10, 10, 20, 35, 55, 120];
 a=polyfit(x,y,m);
 disp('Polynomial coefficients are:'),disp(a')
 disp('Polynomial degree is:'),disp(m)
 x_pol=linspace(min(x),max(x),10);
 y_pol =polyval(a,x_pol);
```

```
plot(x,y,'o', x_pol, y_pol)
xlabel('Concentration of substance'),ylabel('Cancer incidence')
title('Fitting cancer data with polynomial')
grid
end
```

This function file with this function has the name `cancer`. For running, the function requires the value of polynomial degree, m. The first line of `cancer` represents the definition of the function and the next three lines represent function help. After running, the function determines the fitting coefficients with the `polyfit` command, displays the latter and the polynomial degree with the `disp` commands, and plots the fit results together with the original data. If the inputted degree is more than 5 (the number of coefficients in the polynomial is $m + 1$ and it cannot be more than the number of points), the program displays 'Error: Polynomial degree cannot be more than 5', and the run should be repeated with the correct m value.

The Command Window and generated plot for polynomial degree 3 are:

```
>> cancer(3)
Polynomial coefficients are:
  0.0000
 -0.0012
  0.2472
  5.1629
Polynomial degree is:
  3
```

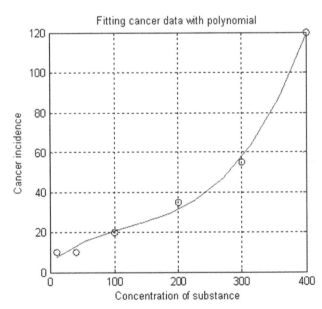

4.5 Questions for self-checking and exercises

1. The user-defined MATLAB® function has in the first line: (a) the function definition; (b) help comments; or (c) definition of global variables?
2. Local variables are: (a) all variables defined, recognized and used only within the function file; (b) input/output parameters only, written in the function definition line; or (c) the variables appearing in the workspace after the function file has completed its work?
3. The `unterp1` function is used: (a) for interpolation only; (b) for extrapolation only; or (c) for both interpolation and extrapolation?
4. In which form should the equation $ae^{-kx} = b$ be written for solution with the aid of the `fzero` function: (a) just as it is written: $b = ae^{-kx}$; (b) $ae^{-kx} - b = 0$; or (c) $\ln(b/a) = -kx$?
5. The function $f(x)$ is given by data points; which MATLAB® function should be used for its integration: (a) `quad`, (b) `trapz` or (c) `fzero`?
6. The expression

$$C = \frac{5}{9}(F - 32)$$

is used to convert degrees Fahrenheit F into degrees Centigrade C. Write the script that can be used as a Fahrenheit to Centigrade converter; display the result to two decimal places.
7. The death time t of spores in a thermal sterilization process is described by the equation

$$t = 3.5 \times 10^{-5}e^{0.083T}$$

where T is temperature in degrees Celcius, and t is time in seconds. Write the script that calculates and displays the death time of the spores at 100, 110, ..., 180 °C. Plot the graph T as function of t, and add axis labels, grid and captions to the graph.
8. The van der Waals gas equation of state is:

$$(P + \frac{an^2}{V^2})(V - nb) = RT$$

where P is pressure in atmospheres, V is volume in liters, T temperature in Kelvin, n is the number of moles, and a and b are constants that should be specified for each gas. Write the function file that calculates the volume of the gas at given P, T, n, a and b; use the `fzero` function to obtain V at $P = 6$ atm, $T = 50$ °C (323.15 K), $n = 2$ mol, $R = 0.08206$ L-atm/mol-K, $a = 5.43$ L^{2}atm/mol^2, $b = 0.0304$ L/mol (water) and initial $V = 5$.

Published by Woodhead Publishing Limited

9. The waist-to-hip ratio (WHR) R = waist/hip is used as an indicator or measure of the state of health of a person; when this ratio is in the range 0.7–0.85 for women and 0.76–0.9 for men, it is ranked as normal; persons with higher WHRs ('apple-shaped' body) face a higher health risk than those with lower WHRs ('pear-shaped' body). Write a function file with waist and hip as input parameters, and R and information about the person's state of health (normal, high health risk, low health risk) as output parameters.

10. New York city population between 1900 and 2000:

Year	1900	1920	1940	1960	1980	2000
People (million)	3.437	5.620	7.455	7.781	7.072	8.008

Write a user-defined function with the table data, and enter into it the `interp1` command with the `'linear'` method to calculate the population within and outside the table range; save the defined function in the function file and use it to calculate the population in 1899, 2010 and 2012.

11. The viscosity of a solution η is 1.75, 1.5, 1.27, 1.3, 1.15 and 1.1 cP at temperatures t of 20, 25, 30, 35, 40 and 45 °C, respectively. Write a user-defined function that describes the graph η(t) together with interpolated values, and add to the graph axis labels, grid, legend and a caption; use the `interp1` command with the 'cubic' method to obtain the viscosity at temperatures within the above range; save the function into a function file and use it to obtain the viscosity at 22.5 °C.

12. The integral of a function $f(x)$ can be calculated by the left Riemann sum equation:

$$I = \Delta x(f_1 + f_2 + \dots + f_{n-1})$$

where n is the number of points in the integration interval $[a,b]$, $\Delta x = (b - a)/(n - 1)$ is the spacing of the points, and f_i is the value of the function at point i = 1, 2, ..., $n - 1$. Calculate by this the bacterial population

$$N = \int_a^b v(t)dt$$

where time t is in hours, the proliferation rate $v(t) = 0.6931 \cdot N_0 \cdot 2^t$, the initial bacterial amount N_0 = 10 cells, and the integration limits are $a = 0$ and $b = 7$ hours. Write and save the program as a function file with input parameters a, b, n, N_0 and output parameter N; run it with $n = 1000$. The resulting N should be given rounded to the nearest integer value.

13. Solve the integral in exercise 12 by the right Riemann sum equation:

$$I = \Delta x(f_2 + f_3 + \dots + f_n)$$

where Δx and f_i have the same sense as in exercise 12. Write and save the program as a function file; run it with $n = 100$, 500 and 1000. Use the function written in exercise 12 to display the results obtained for the left and right Riemann methods. Note: the input and output parameters of the function created should be suitable to solve the problem.

14. Solve the integral in exercise 12 with the quad command. Write and save the program as a function file. Use the functions written in exercises 12 and 13 to display the results obtained with the quad command and with the left and right Riemann methods (for $n = 1000$). Note: the input and output parameters of the function created should be suitable to solve the problem.

15. Bacterial population growth is expressed by the polynomial $p = 10^6 + 10^4 t - 10^3 t^2$, where t is time in hours. Write the script that determines numerically the bacterial growth rate $r = dp/dt$ when $t = 1$ hour. Take a time step equal to 0.01 hour and count the derivatives one step longer than the required t.

16. The decay rate of a certain radioisotope is described by the equation $m = m^0 e^{-kt}$ where t is time, m and m_0 are the isotope amounts at the starting and current time, respectively, and k is the decay constant. Values of m of 500, 250, 125, 63 and 31 µCi were measured on days 0, 20, 40, 60 and 80, respectively. Write a function without parameters that fits these data by the above equation, displays the fitting equation with the found coefficients m_0 and k, and plots the fitting curve and original data.

4.6 Answers to selected exercises

2. (b) Input/output parameters, only written in the function definition line.
4. (b) $ae^{-kx} - b = 0$
6. Enter temperature as Fahrenheit 82
   Temperature 82.00 degrees Fahrenheit is 27.78 degrees Celsius

8.
>> V=Ch4_8(6,323,2,5.43,0.0304)
V =
 3.4495
11.
>>etai=Ch4_11(22.5)
etai =
 1.6224

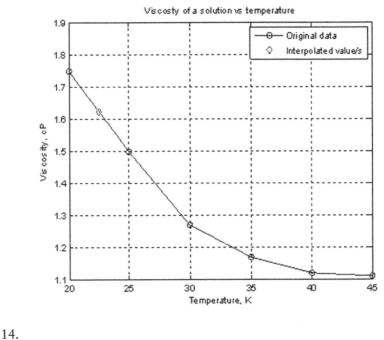

14.
>> [N_right,N_left,N_Simpson]=Ch4_14(0,7,1000,10)
N_right =
 1272
N_left =
 1266
N_Simpson =
 1269
16.
>> Ch4_16
The fitted equation is:
 m=500.80386 e^-0.1499t

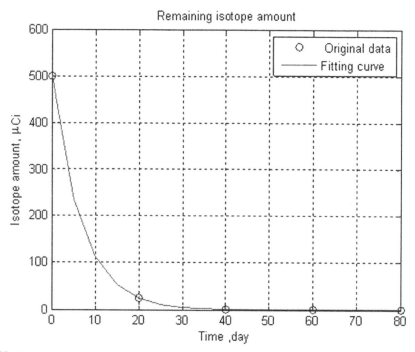

Note

1. The symbol @ is used in MATLAB to denote the so-called anonymous functions, which fall outside the scope of this book.

5

Ordinary and partial differential equation solvers

Differential equations play an important role in science and technology in general and in bioscience and bioengineering in particular. Many processes and phenomena in biokinetics, biomedicine, epidemiology, pharmacology, genetics and other life sciences can be described by such equations. They are often not solvable analytically, in which case a numerical approach is called for, but a universal numerical method does not exist. For this MATLAB® provides tools, called *solvers*, for solutions to two groups of differential equations – ordinary, ODEs, and partial, PDEs. The corresponding groups of solvers, ODEs and PDEPEs, are described in brief below. Basic familiarity with ordinary and partial differential equations is assumed.

5.1   Solving ordinary differential equations with ODE solvers

These solvers are intended for single or multiple ODEs of first order, having the form:

$$\frac{dy_1}{dt} = f_1(t, y_1, y_2, ..., y_n)$$

$$\frac{dy_2}{dt} = f_2(t, y_1, y_2, ..., y_n)$$

...

$$\frac{dy_n}{dt} = f_n(t, t, y_1, y_2, ..., y_n)$$

where n is the number of first-order ODEs, y_1, y_2, ..., y_n are the dependent variables and t is the independent variable; instead of t the variable x can also be used. In the case of high-order ODEs, they should be reduced to first order by creating a set of such equations. For example, the equation $\dfrac{d^2y}{dt_2} + 0.5\left(\dfrac{dy}{dt}\right)^2 y = \sin(t)$ should be rewritten as a set of two first-order ODEs: $\dfrac{dy_1}{dt} = y_2$, $\dfrac{dy_2}{dt} = -0.5y_2^2 y_1 + \sin(t)$. Nevertheless, among practical life science problems, first-order ODEs predominate and thus solution of such equations is discussed further.

Table 5.1 on p. 135 lists the available solvers in MATLAB®, the numerical method each solver uses and the class of problem associated with each solver. These solvers are intended for so-called initial-value problems, IVPs, when the differential equation is solved with any initial value of the function, e.g. $y = 0$ at $t = 0$ for the equation $\dfrac{dy}{dt} = f(t,y)$.

The methods used in those solvers are based on finite differences by which the differential is presented: $\dfrac{dy}{dt} \to \lim\limits_{\Delta t \to 0} \dfrac{\Delta y}{\Delta t} = \dfrac{y_{i+1} - y_i}{t_{i+1} - t_i}$, which is apparently true for very small, but finite (non-zero) distances between the points; i here is the point number in the $[a,b]$ range of the argument t. Giving the first-point value y_0 at t_0 and calculating the value of $\left.\dfrac{dy}{dt}\right|_0 = f(t_0,y_0)$, we get the next y_1. Calculating $\left.\dfrac{dy}{dt}\right|_1 = f(t_1,y_1)$ for the next argument t_1 we get y_2 and the process is repeated until all function values in the range are available. This approach, originally realized by Euler, is used with different improvements and complications, and is used in modern numerical methods such as those of Runge-Kutta, Rosenbrock and Adams.

The class of problems associated with ODE solvers comprise three categories, namely non-stiff, stiff and fully explicit. There is no universally accepted definition of stiffness for the first two categories. A stiff problem is one where the equation contains some terms that lead to such variation as to render it numerically unsolvable even with very small step sizes, and the solution does not converge. In some cases, a rule for categorization of stiff ODEs is that the ratio of the maximal and minimal eigenvalues of the ODE is large, say 100 or more. Unfortunately, this criterion is not always correct and therefore it is impossible to determine in advance what ODE solver should be used for identification of the stiffness of an ODE. In contrast to stiff ODEs, an important feature of a non-stiff equation is that it is stable and the solution converges. Sometimes the stiff category of

Published by Woodhead Publishing Limited

Table 5.1 MATLAB® ODE solvers

Solver name	Numerical method	Sort of problem	Assignment
ode45	Explicit Runge–Kutta method	Non-stiff differential equations	Try first when you do not know which solver is suitable to your non-stiff equation.
ode23	Explicit Runge–Kutta method	Non-stiff differential equations	For non-stiff and moderately stiff problems. Often quicker, but less precise than ode45.
ode113	Adams' method	Non-stiff differential equations	For problems with stringent error tolerances or for solving computationally intensive problems.
ode15s	Numerical differentiation formulas, NDFs (backward differentiation formulas, BDFs)	Stiff differential equations and differential algebraic equations, DAEs	For stiff problem when ode45 is slow. Try first when you do not know which solver is suitable to your stiff equation.
ode23s	Rosenbrock's method	Stiff differential equations	For stiff problem when ode15s is slow.
ode23t	Trapezoidal rule	Moderately stiff differential equations and DAEs	For moderately stiff problems.
ode23tb	Trapezoidal rule/second-order backward differentiation formula, TR/BDF2	Stiff differential equations	For stiff problem, sometimes more effective than ode15s.
ode15i	BDFs	Fully implicit differential equations	For any ODEs given in implicit form $f(t,y,dy/dt) = 0$

Published by Woodhead Publishing Limited

ODEs may be known in advance, for example from some physical reasons (e.g. a fast chemical reaction), but if the category is not known the ode45 solver is recommended first, and the ode15s next.

When an equation cannot be presented in explicit form [e.g. $\frac{dy}{dt} = f(t_1, y_1)$] and has the implicit form $f(t, y, \frac{dy}{dt}) = 0$, the ode15i solver is used. Implicit equations are rarely found in bio-problems and need not be discussed further here.

5.1.1 MATLAB® ODE solver form

The simplest command used in all ODE solvers has the form:

```
[t,y]=odeN(@ode_function,tspan,y0)
```

where odeN is one of the solver names, e.g. ode45 or ode15s; @ode_function is the name of the user-defined function or function file where the equations are written. The function definition line should have the form

```
function dy=function_name(t,y).
```

In the subsequent lines of this file the differential equation(s), presented as described above in Subsection 5.1, should be written in the form

```
dy=[right side of the first ODE;
    right side of the second ODE; ...]
```

tspan – a row vector specifying the integration interval, e.g. [1 12] – specifies a t-interval from 1 to 12; this vector can be given with more than two values for viewing the solution at values written in this vector; for example, [1:2:11 12] means that results lie in the t-range 1–12 and will be displayed at t-values of 1, 3, 5, 7, 9, 11 and 12; the values given in tspan affect the output but not the solution tolerance; as the solver automatically chooses the step for ensuring the tolerance, the default tolerance is 0.000001 absolute units.

y0 is a vector of initial conditions; for example, for the set of two first-order differential equations with initial function values $y_1 = 0$ and $y_2 = 4$, $y_0 = [0\ 4]$. This vector can be given also as column, e.g. $y_0 = [0;4]$.

The odeN solvers can be used without the t and y output arguments; in this case the solver automatically generates a plot.

Published by Woodhead Publishing Limited

5.1.1.1   The steps of ODE solution

The solution of an ODE can be presented in the following steps:

1. As a first step, the differential equation should be given in the form

$$\frac{dy}{dt} = f(t, y), \quad a \leq t \leq b, \quad \text{with } y = y_0 \text{ at } t = t_0$$

   As an example we solve the rate equation of a unimolecular second-order reaction

$$-\frac{d[A]}{dt} = k[A]^2$$

   where $[A]$ is the amount of the reactant A, t is the time of the studied reaction, k is the reaction rate constant – say, $k = 0.5$ mol^{-1} s^{-1}, the initial amount of $A = 0.25$ mol and $t = 0$–10 min.

   This equation has an analytical solution, but here we consider a numerical solution. For this the equation should be rewritten in the form

$$\frac{dy}{dt} = -0.5y^2, \quad 0 \leq t \leq 10 \cdot 60, \quad y = 0.25 \text{ at } t = 0$$

   where $y = [A]$.

2. Now the function file with the user-defined function containing dy/dt as function of t and y should be created. This file must be written in the Editor Window and then saved with a name given to the written function. For our example the function is

   ```
   function dy = ODEfirst(t,y)
   dy = [-0.5*y^2];
   ```

   Therefore, the m-file containing this function is named ODEfirst.

3. In this step the solution method and appropriate ODE solver should be chosen from Table 5.1. In the absence of a specific recommendation, the `ode45` solver for non-stiff problems will serve. The following command should be entered in the Command Window

   ```
   >> [t,y]=ode45(@ODEfirst,[0:100:600], 0.25) % t=0...600 sec, step 100; y0=0.25
   t =
        0
      100
      200
      300
      400
      500
      600
   ```

y =
  0.2500
  0.0185
  0.0096
  0.0065
  0.0049
  0.0039
  0.0033

The process starts at $t = 0$ and ends at $t = 10 \times 60 = 600$ s, the initial value of $y = 0.25$ mol. It yields numbers but does not plot the results. To achieve both, the starting, $t_s = 0$, and finishing, $t_f = 600$, time values only may be given in tspan [which uses a default t-step and adds $y(t)$ points for plotting]; and then the command for plotting should be entered:

>> [t,y]=ode45(@ODEfirst,[0 600], 0.25); % t=0 ...600 s with default step
>>plot(t,y)
>>xlabel('Time, sec'),ylabel('Amount of A, mol')
>>title('Solution of the second order reaction rate equation')
>>grid

The generated plot (Figure 5.1) is:

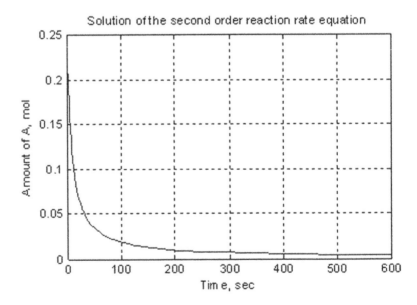

Figure 5.1 Concentration–time dependence for second-order reaction; the solution to $-\dfrac{d[A]}{dt} = k[A]^2$, with $A_0 = 0.25$.

To create a program with all commands included, the function file should be written. For the studied example, this file – named `SecOrdReaction` – reads as follows:

```
function t_A=SecOrdReaction(ts,tf,y0,n)
% Solution of the ODE for second order reaction rate
% t – time, sec; y – amount of the A-reactant, mol
%   tspan=[0 600];y0=0.25;
%   n- number of points for displaying
% To run:>> t_A=SecOrdReaction(0,600,0.25,5)
close all % closes all previously plotted figures
tspan=[ts,tf];
[t,y]=ode45(@ODEfirst, tspan, y0);
plot(t,y)
xlabel('Time, sec'),ylabel('Amount of A, mol')
title('Solution of the second order reaction rate equation')
grid
t_n=linspace(ts,tf,n);% n t points for displaying
y_n=interp1(t,y,t_n,'spline');% n y points for displaying
t_A=[t_n' y_n'];
function dy=ODEfirst(t,y)
dy=-0.5*y^2;
```

The input parameters in the written function are: tspan – time range, y_0 – initial value of y and n – the number of points for displaying results, which should be greater than 2 but below the length of the vector t. The output parameter is the two-column matrix t_A with t in the first column and the amounts of A in the second. The n time-values t_n are generated by the `linspace` command, and the n concentrations y_n interpolated by the `interp1` command; this is necessary because the number of t,y values calculated by the `ode45` command and the desired number of these values can differ.

To run this file the following command should be typed and entered in the Command Window:

```
>>t_A=SecOrdReaction(0,600,0.25,5)
   t_A =
        0              0.2500
   150.0000           0.0127
   300.0000           0.0065
   450.0000           0.0044
   600.0000           0.0033
```

The graph $[A] = f(t)$ is also plotted by the `SecOrdReaction` function.

5.1.1.2 Extended form of the ODE solvers

When ODE solver commands are used for equations with parameters (e.g. in the discussed example the reaction rate constant k is a parameter), a more complicated form is necessary:

```
[t,y]=odeN(@ode_function,tspan,y0,[],
        param_name1,param_name2,...)
```

where [ ] is an empty vector; in the general case it is the place for various options that change the process of integration;[1] in most cases, default values are used which yield satisfactory solutions and we do not use these options. param_name1, param_name2,... are the names of the arguments that we intend to write into the ode_function.

If the parameters are named in the odeN solver, they should also be written into the function.

For example, the SecOrdReaction file should be modified to introduce the coefficient k as an arbitrary parameter:

```
function t_A=SecOrdReaction(k,ts,tf,y0,n)
% Solution of the ODE for second order reaction rate
% k – reaction rate constant, 1/(mole sec)
% t – time, sec; y – amount of the A – reactant, mol
% tspan=[0 600];y0=0.25;
% n – number of points for sisplaying
% To run:>> t_A=SecOrdReaction(0.5,0,600,0.25,5)
close all % closes all previously plotted figures
tspan=[ts,tf];
[t,y]=ode45(@ODEfirst, tspan, y0,[],k);
plot(t,y)
xlabel('Time, sec'),ylabel('Amount of A, mol')
title('Solution of the second order reaction rate equation')
grid
t_n=linspace(ts,tf,n);% n time points for displaying
y_n=interp1(t,y,t_n,'spline');% n concentr. points for displaying
t_A=[t_n' y_n'];
function dy=ODEfirst(t,y,k)
dy=-k*y^2;
```

And in the Command Window the following command should be typed and entered:

```
>> t_A=SecOrdReaction(0.5,0,600,0.25,5)
```

The results are identical to those discussed above.

The advantage of the form with parameters is its greater universality; for example, the `SecOrdReaction` function with parameter k can be used for any second-order reaction with previously obtained reaction rate constant.

5.1.2 Application examples

The commands in the examples below are written mostly as functions with parameters. The single help line in these functions consists of a command that should be typed in the Command Window. In the general case the explanations for input and output parameters are conveniently introduced as described in the preceding chapter.

5.1.2.1 Michaelis–Menten kinetics

The enzyme-catalyzed reactions $E+S\rightleftharpoons ES\rightarrow E+P$ studied by Michaelis and Menten are described by the following set of four ODEs:

$$\frac{d[S]}{dt} = k_{1r}[ES] - k_1[S][E]$$

$$\frac{d[E]}{dt} = (k_{1r} + k_2)[ES] - k_1[S][E]$$

$$\frac{d[ES]}{dt} = k_1[S][E] - (k_{1r} + k_2)[ES]$$

$$\frac{d[P]}{dt} = k_2[ES]$$

where $[S]$, $[E]$, $[ES]$ and $[P]$ are the amounts of substrate, free enzyme, substrate-bound enzyme and product, respectively, k_1 and k_{1r} are the direct and reverse reaction rate constants in the $E+S\rightleftharpoons ES$ reaction, and k_2 is the rate constant of the $ES\rightarrow P$ reaction.

<u>Problem:</u> Solve the set of ODEs in the time range 0–6 s with initial $[S] = 8$ mol, $[E] = 4$ mol, $[ES] = 0$ and $[P] = 0$. Reaction rate constants are: $k_1 = 2$ mol^{-1} s^{-1}, $k_{1r} = 1$ s^{-1} and $k_2 = 1.5$ s^{-1}. Plot the component amounts as a function of time, and display the data obtained.

Formulate the set of equations in the form suitable for the ODE solver:

$$\frac{dy_1}{dt} = k_{1r}y_3 - k_1 y_1 y_2$$

$$\frac{dy_2}{dt} = (k_{1r} + k_2)y_3 - k_1 y_1 y_2$$

$$\frac{dy_3}{dt} = k_1 y_1 y_2 - (k_{1r} + k_2)y_3$$

$$\frac{dy_4}{dt} = k_2 y_3$$

at $0 \leqslant t \leqslant 6$ with $k_1 = 2$, $k_{1r} = 1$, $k_2 = 1.5$ and initial $y_1 = 8$, $y_2 = 4$, $y_3 = 0$ and $y_4 = 0$.

Now, the ODE solver should be chosen. In the absence of preliminary reasons, the `ode45` solver should be adopted.

The function file for the solution is

```
function t_S_E_ES_P=MichaelisMenten(k1,kr1,k2,ts,tf,S0,E0,ES0,P0,n)
% To run >>t_S_E_ES_P=MichaelisMenten(2,1,1.5,0,6,8,4,0,0,10)
tspan=[ts tf];
[t,y]=ode45(@MM,tspan,[S0,E0,ES0,P0],[],k1,kr1,k2);
plot(t,y);
title('Michaelis-Menten Enzyme Kinetics: S<=>E->SE->E+P')
xlabel('Time, sec'),ylabel('S, E, ES, P amounts, mol')
legend('S','E','SE','P')
grid
t_n=linspace(ts,tf,n);% n time points for displaying
y_n=interp1(t,y,t_n,'spline');% n concentrations for displaying
t_S_E_ES_P=[t_n' y_n];
function dc=MM(t,y,k1,kr1,k2)
dc=[kr1*y(3)-k1*y(1)*y(2);
    (kr1+k2)*y(3)-k1*y(1)*y(2);
    k1*y(1)*y(2)-(kr1+k2)*y(3);
    k2*y(3)];
```

The input arguments in the `MichaelisMenten` function: k1, kr1, k2, S0, E0, ES0 and P0 correspond to k_1, k_{1r}, k_2, $[S]_0$, $[E]_0$, $[ES]_0$ and $[P]_0$ in the original set; the tspan is a two-element vector with starting and end values of t; n, as in the preceding example, is the number of t,y-rows to be displayed.

The t_S_E_ES_P output argument is a five-row matrix with t and $[S]$, $[E]$, $[ES]$ and $[P]$ as the component amounts. As in the preceding example, the `linspace` and `interp1` commands are used to display the n-rows of desired times t_n and accordingly the concentrations y_n.

Published by Woodhead Publishing Limited

To run the function in the Command Window, the following command should be typed and entered:

```
>> t_S_E_ES_P=MichaelisMenten(2,1,1.5,0,6,8,4,0,0,10)
t_S_E_ES_P =
```

0	8.0000	4.0000	0	0
0.6667	2.4781	1.2163	2.7837	2.7382
1.3333	0.8737	2.0208	1.9792	5.1471
2.0000	0.2623	2.9226	1.0774	6.6604
2.6667	0.0898	3.5051	0.4949	7.4154
3.3333	0.0346	3.7859	0.2141	7.7513
4.0000	0.0140	3.9092	0.0908	7.8952
4.6667	0.0058	3.9618	0.0382	7.9560
5.3333	0.0024	3.9840	0.0160	7.9815
6.0000	0.0010	3.9933	0.0067	7.9923

The resulting graph is:

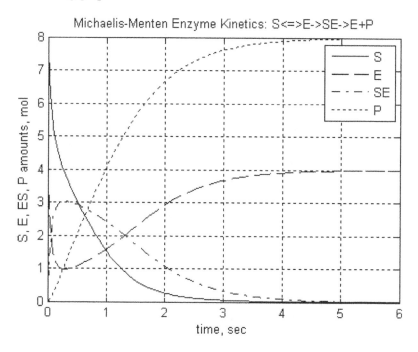

5.1.2.2 Chemostat

A chemostat[2] is a container of volume V with constant culture medium inflow F and fresh nutrient concentration R. The differential equations for the concentration of bacteria N and that of nutrient S are:

$$\frac{d[N]}{dt} = [N]\left(\frac{r[S]}{a+[S]} - D\right)$$

$$\frac{d[S]}{dt} = D([R]-[S]) - \frac{1}{\gamma}[N]\frac{r[S]}{a+[S]}$$

where $D = F/V$ is the dilution rate, r is the maximal growth rate, a is the half-saturation constant and γ is growth yield; square brackets [ ] denote concentration.

<u>Problem:</u> Solve the set of ODEs in the time range 0–10 s with $r = 1.35$ h^{-1}, $a = 0.004$ g/L, $D = 0.25$ h^{-1}, $[R] = 6$ g/L, $\gamma = 0.23$, $[N]_0 = 0.1$ g/L and $[S]_0 = 6$ g/L. Plot the nutrient and bacteria concentrations as a function of time, and display the data obtained.

Formulate the set of differential equations in the form suitable for the ODE solvers:

$$\frac{dy_1}{dt} = y_1\left[\frac{r\, y_2}{a+y_2} - D\right]$$

$$\frac{dy_2}{dt} = D(R-y_2) - \frac{1}{\gamma}y_1\frac{r\, y_2}{a+y_2}$$

where y_1 is [N] and y_2 is [S], t is varied between $0 \leqslant t \leqslant 6$, $k_1 = 2$, $k_{1r} = 1$, $k_2 = 1.5$ and the initial $y_1 = 6$ and $y_2 = 0.1$.

The ODE solver should now be chosen; as the chemostat problem is stiff, the `ode15s` solver should be used.

The `chemostat` function file that solves the problem is:

```
function t_N_C = chemostat(r,a,gam,D,R,ts,tf,N0,S0,n)
%To run:>> t_N_C = chemostat(1.35,0.004,0.23,0.25,6,0,10,0.1,6,10)
close all
tspan = [ts tf];
[t,y] = ode15s(@chem_equations,tspan,[N0;S0],[],r,a,gam,D,R);
plot(t,y)
xlabel('Time, hour')
ylabel('Amount of Nutrient and Microorganisms, g/L')
legend('Nutrient','Microorganisms')
grid
t_n = linspace(ts,tf,n);% n time points for displaying
y_n = interp1(t,y,t_n,'spline');% n concentrations for displaying
t_N_C = [t_n' y_n];
function dy = chem_equations(t,y,r,a,gam,D,R)
```

```
%y(1) is N;y(2) is S
ef = r*y(2)(a+y(2));
dy = [y(1)*(ef-D);D*(R-y(2))-ef*y(1)/gam];
```

The input arguments in the `chemostat` function, r, a, gam, D and R, correspond to r, a, y, D and R in the original equation set; the tspan is a two-element vector with starting and end values of t; and n is the number of points for display.

The t_N_S output argument is a three-row matrix with t and concentrations [N] and [S]. The `linspace` and `interp1` commands are used as previously to display the n rows of the time and concentration values. For brevity, the function help was reduced to a single line containing an example of function call from the Command Window; such a reduction will apply later in other examples.

To run the function in the Command Window, the following command should be typed and entered

```
>> t_N_C = chemostat(1.35,0.004,0.23,0.25,6,0,10,0.1,6,10)
t_N_C =
```

0	0.1000	6.0000
0.4081	0.1570	5.7100
2.1061	1.0183	1.8295
2.4228	1.4320	0.0114
2.4265	1.4342	0.0013
2.4286	1.4343	0.0009
2.4320	1.4342	0.0009
2.4456	1.4341	0.0009
3.5507	1.4212	0.0009
10.0000	1.3880	0.0009

The resulting plot is shown at the top of p. 146.

5.1.2.3 *Predator–prey Lotka–Volterra model*

The Lotka–Volterra is the simplest model for competition between prey, x, and predators, y. With certain simplifications, the model can be represented by the two differential equations:

$$\frac{dx}{dt} = k_1 x - k_2 xy$$

$$\frac{dy}{dt} = -k_3 y + k_4 xy$$

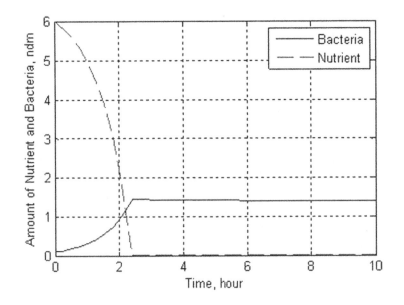

where k_1 and k_3 are the growth rate of x (prey, e.g. rabbits) and y (predators, e.g. foxes) species, respectively, and k_2 and k_4 are coefficients characterizing the interactions between the species.

Problem: Solve the set of ODEs in the time range 0–15 (dimensionless) with initial $x = 5000$ (rabbits) and $y = 100$ (foxes). The growth rate constants are: $k_1 = 2$, $k_2 = 0.01$, $k_3 = 0.8$, $k_4 = 0.0002$. Plot the species populations as a function of time, and display the data obtained.

The set of differential equations in the form suitable for ODE solvers is:

$$\frac{dy_1}{dt} = k_1 y_1 - k_2 y_1 y_2$$

$$\frac{dy_2}{dt} = -k_3 y_2 + k_4 y_1 y_2$$

where y_1 is x and y_2 is y; $0 \leqslant t \leqslant 5$, $k_1 = 2$, $k_2 = 1.5$, $k_3 = 0.8$ and $k_4 = 0.0002$, and initial $y_1 = 5000$ and $y_2 = 100$.

In the absence of a specific recommendation, the ode45 solver should be used. The function file for the set solution is

    function t_prey_pred = PredPrey(k1,k2,k3,k4,ts,tf,x0,y0,n)
    %To run: >> t_t_prey_pred = PredPrey(2,0.01,0.8,0.0002,0,15,5000,
    100,15)
    close all

```
[t,y] = ode15s(@LotVol,tspan,[x0;y0],[],k1,k2,k3,k4);
plot(t,y(:,1),'-',t,y(:,2),'--')
xlabel('Time, hour')
ylabel('Amount of predators and preys vs time')
legend('Prey – rabbit','Predator – fox')
grid
t_n = linspace(ts,tf,n);% n time points for displaying
y_n = interp1(t,y,t_n,'spline');% n concentrations for displaying
t_prey_pred = [t_n' y_n];
function dy = LotVol(t,y,k1,k2,k3,k4)
%y(1) is x; y(2) is y
dy = [k1*y(1)-k2*y(1)*y(2);-k3*y(2)+k4*y(1)*y(2)];
```

The input arguments, k1, k2, k3, k4 and x0 and y0 are, respectively, the same as the k_1, k_2, k_3, k_4 and y_1 and y_2 values at $t = 0$ in the original set; the tspan is a two-element vector with starting and end t values; and n is the number of points for display. The t_prey_pred output argument is a three-column matrix with the t and population values of prey and predator. Here as in the preceding examples the n rows of t and y values are obtained by the `linspace` and `interp1` commands.

After running this function in the Command Window the following results are displayed and plotted:

```
>> t_prey_pred = PredPrey(2,0.01,0.8,0.0002,0,16,5000,100,15)
t_prey_pred =
   1.0e+003 *
        0           5.0000        0.1000
   0.0004           7.2701        0.1189
   0.0013           8.3263        0.2799
   0.0022           2.5578        0.3399
   0.0032           1.1982        0.2061
   0.0045           2.2467        0.1102
   0.0062           9.4895        0.1968
   0.0071           3.8686        0.3564
   0.0080           1.4043        0.2659
   0.0092           1.4701        0.1400
   0.0111           8.3298        0.1367
   0.0122           5.0035        0.3521
   0.0131           1.5950        0.2897
   0.0142           1.2852        0.1607
   0.0160           6.0432        0.1060
```

Published by Woodhead Publishing Limited

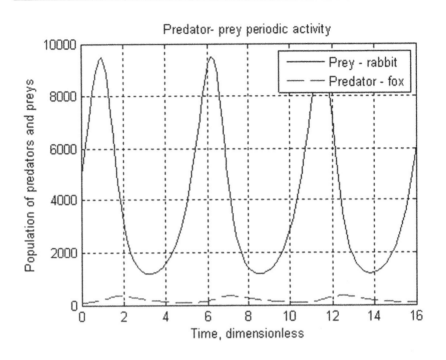

5.1.2.4 Malthus–Verhulst population model

The Malthus–Verhulst ODE is frequently used in population dynamics:

$$\frac{dN}{dt} = rN\left(1 - \frac{N}{K}\right)$$

where N is the population size, r is the population growth/decline rate coefficient and K is the maximum population the environment can support. This equation has an analytical solution:

$$N = \frac{K}{1 + CKe^{-rt}}$$

in which $C = 1/N_{ref} - 1/K$, and N_{ref} is a reference point of the population data.

Problem: Solve numerically the Malthus–Verhulst differential equation with initial $N_0 = 8.5$ million people, and constants $K = 395.8$ and $r = 1.354$ (values drawn from USA population statistics). Time is given as a centered and scaled parameter $(t - 1890)/62.9$ with $t = 1790, 1810, ..., 2050$; calculate N also with the analytical solution equation when $N_{ref} = 62.9$, at $(t - 1890)/62.9 = 0$. Plot the numerical and theoretical solution data in the same graph.

Published by Woodhead Publishing Limited

The solution is realized in the following steps:

1. The differential equation is solved; for this it should be rewritten in the form suitable for ODE solvers:

$$\frac{dy}{dt} = ry\left(1 - \frac{y}{K}\right)$$

The time range is $(1790-1890)/62.9 \ldots (2050-1890)/62.9$, and initial $y_0 = 8.5$.

The `ode45` solver is chosen for this equation.

2. The theoretical solution is obtained by the analytical equation at the time values defined in the previous step.

3. Results are displayed and presented in the graph.

Create for solution the `Population` function file:

```
function t_numeric_theoretic = Population(r,K,N0,tspan)
% To run: >>t_numeric_theoretic = Population(1.354,395.8,8.5,[1790:
%20:2050])
close all
t1 = (tspan-1890)./62.9;
[t1,y] = ode45(@MaltVerh,t1,N0,[],r,K);
tspan = tspan'; % row to column for including in t_numeric_theoretic
y_theor = K./(1+K*(1/62.9-1/K)*exp(-r*t1);
t_numeric_theoretic = [tspan,y,y_theor];
plot(tspan,y,tspan,y_theor,'o')
xlabel('Time, year'), ylabel('Population, million')
title('Malthus-Verhulst population model')
legend('Numeric solution','Theoretical solution','Location','Best')
grid
function dy = MaltVerh(t1,y,r,K)
dy = r*y*(1-y/K);
```

The input arguments for the `Population` function, r, K and N0, are the same as the r, K and y0; tspan is here a row vector with years given with the `colon` (:) operator – [1890:20:2050].

The output argument is the `t_numeric_theoretic` three-column matrix with time in the first column, the population obtained by the `ode45 solver` in the second and the population calculated by the theoretical expression in the third.

Published by Woodhead Publishing Limited

After running this function in the Command Window, the following results are displayed and plotted:

>> t_numeric_theoretic=Population(1.354,395.8,8.5,[1790:20:2050])
t_numeric_theoretic =
1.0e+003 *

1.7900	0.0085	0.0081
1.8100	0.0130	0.0124
1.8300	0.0198	0.0189
1.8500	0.0298	0.0285
1.8700	0.0442	0.0425
1.8900	0.0645	0.0620
1.9100	0.0916	0.0884
1.9300	0.1258	0.1219
1.9500	0.1658	0.1614
1.9700	0.2086	0.2042
1.9900	0.2505	0.2463
2.0100	0.2879	0.2843
2.0300	0.3186	0.3158
2.0500	0.3422	0.3401

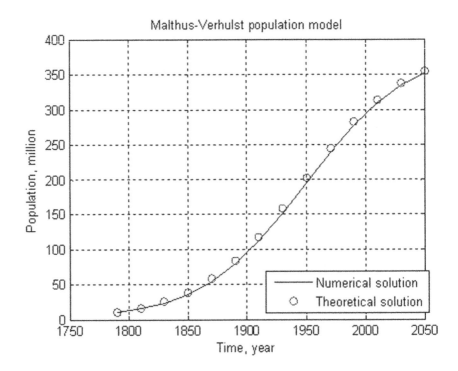

5.2   Solving partial differential equations with the PDE solver

Given the relevance of PDEs in certain biological disciplines, a short introduction to the MATLAB® PDE solver is given here. This solver is designed for the solution of spatially one-dimensional (1D) PDEs and can be used for a single equation or a set that can be presented in the standard form:

$$c\left(x,t,u,\frac{\partial u}{\partial t}\right)\frac{\partial u}{\partial t} = x^{-m}\frac{\partial}{\partial x}\left(x^{m}f\left(x,t,u,\frac{\partial u}{\partial x}\right)\right)+s\left(x,t,u,\frac{\partial u}{\partial x}\right)$$

where time t is defined in the range t_0 (initial) to t_f (final) and the coordinate x between $x = a$ and b; m can be 0, 1 or 2 corresponding to slab, cylindrical or spherical symmetry, respectively, the first case representing the elliptical and the other two the parabolic PDE type. Accordingly, the MATLAB® PDE solver is suitable for elliptical or parabolic PDEs only; $f\left(x,t,u,\frac{\partial u}{\partial x}\right)$ is called a flux term, and $s\left(x,t,u,\frac{\partial u}{\partial x}\right)$ a source term.

The initial condition for this equation is

$$u(x,t_0) = u_0(x)$$

and the boundary condition at both points $x = a$ and $x = b$ is

$$p(x,t,u)+g(x,t)f\left(x,t,u,\frac{\partial u}{\partial x}\right)=0$$

Note that in this equation the same f function is used as previously in the PDE.

In the case of two or more PDEs the c, f, s and q terms in the standard equation can be written as column vectors; in this case element-by-element operations should be applied.

The functions in these equations are subject to certain restrictions which will be explained below. More detailed information is obtainable by typing
>>doc pdepe
or by using MATLAB® Help. The pdepe is the solver command for solution of PDEs.

5.2.1   The steps of PDE solution

Solution of a single PDE goes through the following steps:
1.   First, the equation should be presented in the accepted form.

For example, consider the simple diffusion equation for the temporal change of a component concentration:

$$\frac{\partial u}{\partial t} = D\frac{\partial^2 u}{\partial x^2}$$

All variables in this equation are given in dimensionless units, x ranging from 0 to 1, and t from 0 to 0.4. The initial ($t = 0$) concentration is $u = 1$ at x between 0.45 and 0.55 and $u = 0$ elsewhere. The boundary conditions are

$$u(0,t) = 0, \qquad u(1,t) = 0$$

The dimensionless diffusion coefficient D is taken as 0.3.
Comparing this equation with the standard form, we obtain:

$$c\left(x,t,u,\frac{\partial u}{\partial t}\right) = 1,$$

$$m = 0,$$

$$f\left(x,t,u,\frac{\partial u}{\partial x}\right) = D\frac{\partial u}{\partial x},$$

$$s\left(x,t,u,\frac{\partial u}{\partial x}\right) = 0.$$

Thus, the diffusion equation in standard form can be rewritten as:

$$\frac{\partial u}{\partial t} = \frac{\partial}{\partial x}\left(D\frac{\partial u}{\partial x}\right)$$

2. The initial and boundary conditions should be presented in standard form.
    Accordingly they give:

$$u(x,0) = \begin{cases} 1, & 0.45 \le x \le 0.55 \\ 0, & \text{otherwise} \end{cases}$$

and

$$u(0,t) + 0 \cdot D\frac{\partial u(0,t)}{\partial x} = 0$$

$$u(1,t) + 0 \cdot D\frac{\partial u(1,t)}{\partial x} = 0$$

respectively.
To obtain these equations, the following conventions were used:

$$p(x,t,u) = u$$

$$g(x,t) = 1$$

$$f\left(x,t,u,\frac{\partial u}{\partial x}\right) = D\frac{\partial u}{\partial x}$$

3. At this stage the `pdepe` solver and the functions with the solving PDE, with initial conditions, and boundary conditions should be used. The `pdepe` solver command has the form

    ```
    sol = pdepe(m,@pde_function,@pde_ini_cond,@pde_bon_
    cond,x_mesh,tspan)
    ```

    where m is 0, 1 or 2 as above. The `pde_function` is the name of the function where the solving differential equation should be written; its definition line reads

    ```
    function [c,f,s] = pdefun(x,t,u,DuDx)
    ```

    the arguments `x`, `t`, `u`, `DuDx`, `c`, `f` and `s` being the same x, t, u, $\partial u/\partial x$, c, f and s as in the standard form; the derivative $\partial u/\partial x$ in standard expression should be written `DuDx`. The command `pde_ini_cond` is the name of the function with the initial condition of the solving equation; its definition line reads

    ```
    function u0 = pde_ini_cond(x)
    ```

    u0 being the vector $u(x)$ value at $t = 0$. `pde_bon_cond` is the name of the function with the boundary conditions; its definition line reads

    ```
    function [pa,qa,pb,qb] = pde_bon_cond(xa,ua,xb,ub,t)
    ```

    xa = a and xb = b being the boundary points, ua and ub the u values at these points, pa, qa, pb and qb being the same as in the standard form, and representing the p and q values at a and b; x_mesh is a vector of the x-coordinates of the points at which the solution is sought for each time value contained in the tspan; the serial number of each such point and the respective coordinate should be given at the appropriate place; xmesh values should be written in ascending order from a to b; tspan is a vector of the time points as above; the t values should likewise be written in ascending order.

Published by Woodhead Publishing Limited

The output argument `sol` is a 3D array comprising k 2D arrays (see Section 2.2) with i rows and j columns each. Elements in 3D arrays are numbered analogously to those in 2D arrays; for example, `sol(1,3,2)` denotes a term of the second array located in the first row of the third column, and `sol(:,3,1)` denotes in the first array all rows of the third column. In the pdepe solver `sol(i,j,k)` denotes defined u_k at the t_i time- and x_j coordinate points; for example, sol(:,:,1) is an array whose rows contain the u values at all x-coordinates for each of the given times; $k = 1$ for a single PDE, and $k = 1$ and 2 for a set of two PDEs.

To solve the diffusion equation of our example, the program written as a function and saved in the file with name `mydiffusion` is:

```
function mydiffusion(m,n_x,n_t)
% To run: >> mydiffusion(0,100,100)
close all
xmesh = linspace(0,1, n_x);tspan = linspace(0,0.4, n_t);
sol = pdepe(m,@mypde,@mydif_ic,@mydif_bc,xmesh,tspan);
u = sol(:,:,1);
[X,T] = meshgrid(xmesh,tspan);
figure
mesh(X,T,u)
xlabel('Coordinate'),ylabel('Time'), zlabel('Concentration')
title('Concentration as function of x and t')
figure
c = contour(X,u,T,[0.001 0.005 0.05]);
clabel(c)
xlabel('Coordinate'), ylabel('Concentration')
title('Concentration as function of x at constant t')
grid
function  [c,f,s] = mypde(x,t,u,DuDx)
% PDE for solution
c = 1;
D = 0.1;
f = D*DuDx;
s = 0;
function u0 = mydif_ic (x)
% initial condition
if x< = 0.45||x> = 0.55
   u0 = 0;
else
   u0 = 1;
end
```

```
function [pa,qa,pb,qb] = mydif_bc(xa,ua,xb,ub,t)
% boundary conditions
pa = ua;pb = ub;
qa = 0;qb = 0;
```

The `mydiffusion` function is written without output arguments; its input arguments are: m, as above, and n_x and n_t are the serial numbers of the operative x and t points of the solution. This function contains three subfunctions: `mypde` comprising the diffusion equation, `mydif_ic` defining the initial condition and `mydif_bc` defining the boundary conditions. The x_mesh and tspan vectors are created with two `linspace` functions in which the n_x and n_t are inputted by the `mydiffusion`. The numerical results are stored in the `sol` 3D array, rewritten as a 2D matrix u for subsequent use in the graphical commands. The `mesh` command is used to generate the 3D mesh plot $u(x,t)$ and the `contour` and `clabel` commands – to generate the 2D plot $u(x)$ with the iso-time lines defined in the `contour` as the vector of the selected t values.

After running this function in the Command Window:

```
>> mydiffusion(0,100,100)
```

the following two plots are generated (Figure 5.2, p. 156). The `clabel` command fortuitously superimposes the labels 0.001 and 0.05 on one another and they were separated with the plot editor (see subsection 3.1.3.2) so as not to overburden the program.

In the case of multiple PDEs, the c, f and s functions in the general PDE form and the u_0, p and q in the initial and boundary value equations are column vectors.

5.2.2 Application examples

5.2.2.1 *Reaction–diffusion two-equation system*

For bio-mathematical models, a set of two reaction–diffusion equations is used:

$$\frac{\partial u}{\partial t} = \frac{1}{2}\frac{\partial^2 u}{\partial x^2} + \frac{1}{1+v^2}$$

$$\frac{\partial v}{\partial t} = \frac{1}{2}\frac{\partial^2 v}{\partial x^2} + \frac{1}{1+u^2}$$

where u and v are concentrations of the participating reagents; all parameters are dimensionless, x varying from 0 to 1 and t from 0 to 0.3.

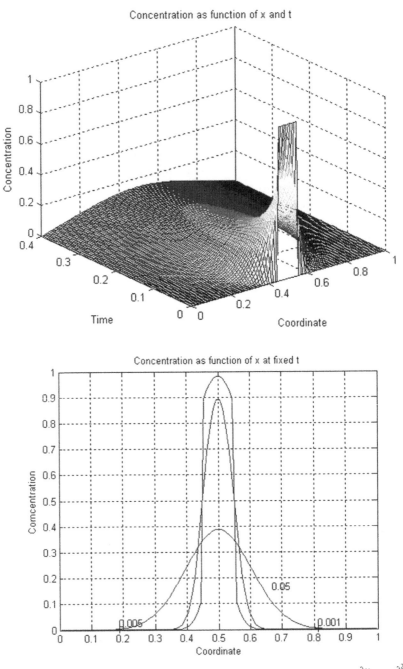

Figure 5.2 Concentration–coordinate–time dependences; the solution to $\dfrac{\partial u}{\partial t} = D\dfrac{\partial^2 u}{\partial x^2}$.

The level values replaced by the Plot Editor.

The initial conditions are assigned as

$$u(x,0) = 1 + \frac{1}{2}\cos(2\pi x)$$

$$v(x,0) = 3$$

and the boundary conditions as

$$\frac{\partial u(0,t)}{\partial x} = \frac{\partial u(1,t)}{\partial x} = 0$$

$$\frac{\partial v(0,t)}{\partial x} = \frac{\partial v(1,t)}{\partial x} = 0$$

Problem: Solve the set of two PDEs with given initial and boundary conditions and present the changes in the concentrations with time t and along coordinate x.

Rewrite first the set in the standard form

$$\begin{bmatrix} 1 \\ 1 \end{bmatrix} \cdot * \frac{\partial}{\partial t}\begin{bmatrix} u_1 \\ u_2 \end{bmatrix} = \frac{\partial}{\partial x}\begin{bmatrix} \dfrac{1}{2}\dfrac{\partial u_1}{\partial x} \\ \dfrac{1}{2}\dfrac{\partial u_2}{\partial x} \end{bmatrix} + \begin{bmatrix} \dfrac{1}{1+u_2^2} \\ \dfrac{1}{1+u_1^2} \end{bmatrix}$$

where u_1 is u and u_2 is v; the m argument is 0 and efficients with x^0 and x^{-0} are absent.

Note: Element-wise multiplication is used here (see subsection 2.2.2), and thus the left-hand part is apparently the vector $\begin{bmatrix} \dfrac{\partial u_1}{\partial t} \\ \dfrac{\partial u_2}{\partial t} \end{bmatrix}$.

The initial conditions in the standard form are

$$\begin{bmatrix} u_1(x,0) \\ u_2(x,0) \end{bmatrix} = \begin{bmatrix} 1 + \dfrac{1}{2}\cos(2\pi x) \\ 3 \end{bmatrix}$$

and the boundary conditions at $x = a$ and $x = b$ are:

$$\begin{bmatrix} 0 \\ 0 \end{bmatrix} \cdot * \begin{bmatrix} u_1 \\ u_2 \end{bmatrix}_a + \begin{bmatrix} 1 \\ 1 \end{bmatrix} \cdot * \begin{bmatrix} \dfrac{1}{2}\dfrac{\partial u_1}{\partial x} \\ \dfrac{1}{2}\dfrac{\partial u_2}{\partial x} \end{bmatrix}_a = \begin{bmatrix} 0 \\ 0 \end{bmatrix}$$

$$\begin{bmatrix} 0 \\ 0 \end{bmatrix} \cdot * \begin{bmatrix} u_1 \\ u_2 \end{bmatrix}_b + \begin{bmatrix} 1 \\ 1 \end{bmatrix} \cdot * \begin{bmatrix} \dfrac{1}{2}\dfrac{\partial u_1}{\partial x} \\ \dfrac{1}{2}\dfrac{\partial u_2}{\partial x} \end{bmatrix}_b = \begin{bmatrix} 0 \\ 0 \end{bmatrix}$$

Now write the `ReactDiff` function file in the form of function without arguments:

```
function ReactDiff
% To run >>ReactDiff
close all
m = 0;
n_x = 20;
n_t = 100;
xmesh = linspace(0,1,n_x);tspan = linspace(0,0.3,100);
sol = pdepe(m,@ReactDiffPDE,@ReactDiff_ic,@ReactDiff_bc,xmesh,tspan);
u = sol(:,:,1);
v = sol(:,:,2);
subplot(1,2,1)
[X,T] = meshgrid(xmesh,tspan);
surf(X,T,u)
xlabel('Coordinate'),ylabel('Time'),zlabel('Concentration u')
title('Concentration of the first reagent')
axis square
subplot(1,2,2)
surf(X,T,v)
xlabel('Coordinate'),ylabel('Time'),zlabel('Concentartion v')
title('Concentration of the second reagent')
axis square
function [c,f,s] = ReactDiffPDE(x,t,u,DuDx)
% PDEs for solution
c = [1;1];
f = DuDx/2;
s = [1/(1+u(2)^2);1/(1+u(1)^2)];
function u0 = ReactDiff_ic(x)
% initial conditions
u0 = [1+0.5*cos(2*pi*x);3];
function [pa,qa,pb,qb] = ReactDiff_bc(xa,ua,xb,ub,t)
% boundary conditions
pa = [0;0];qa = [1;1];
pb = [0;0];qb = [1;1];
```

The `ReactDiff` function is written without output arguments, and its input arguments are: m as above, and n_x and n_t are the numbers of x and t points required. As in the single-PDE example, three sub-functions are involved: the ReactDiffPDE comprising the two reaction–diffusion equations written in the two-row matrices c, f and s, ReactDiff_ic defining the initial conditions in the two-row matrix u_0, and `mydif_bc` defining the conditions at $x = a$ and $x = b$. The `xmesh` and `tspan` vectors are created with two `linspace` functions. The numerical results are stored in the `sol` 3D array, rewritten as 2D matrices u and v for subsequent use in the graphical commands. Two subplot commands are used to place two graphs on the same page: `surf` for generating the 3D surface plots $u(x,t)$ and $v(x,t)$, and `axis square` for rendering the current axis box square in shape.

After running this function in the Command Window the following plots are generated:

>> ReactDiff

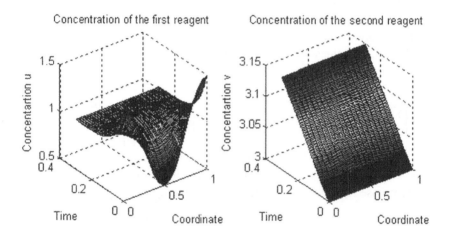

5.2.2.2 Model for the first steps of tumor-related angiogenesis

Tumor-related angiogenesis is an important subject of current oncology research. The first steps of this phenomenon are described by a set of two PDEs:[3]

$$\frac{\partial n}{\partial t} = \frac{\partial}{\partial x}\left(d\frac{\partial n}{\partial x} - a \cdot n \frac{\partial c}{\partial x} \right) + s \cdot r \cdot n \cdot (N - n)$$

$$\frac{\partial c}{\partial t} = \frac{\partial}{\partial x}\left(\frac{\partial c}{\partial x} \right) + s\left(\frac{n}{n+1} - c \right)$$

where N is the total cell population, n is the concentration of the capillary endothelial cells (ECs), c is that of fibronectin, and d, a, s and r are experimentally determined constants. In dimensionless form these coefficients were obtained as $N = 1$, $d = 0.003$, $a = 3.8$, $s = 3$ and $r = 0.88$.

Published by Woodhead Publishing Limited

The initial conditions have small alterations along x and are described by the step functions

$$n(x,0) = \begin{cases} 1.05, & 0.3 \le x \le 0.6 \\ 1, & \text{otherwise} \end{cases}$$

$$c(x,0) = \begin{cases} 0.5025, & 0.3 \le x \le 0.6 \\ 0.5, & \text{otherwise} \end{cases}$$

The boundary conditions are

$$\frac{\partial n(0,t)}{\partial x} = \frac{\partial n(1,t)}{\partial x} = 0$$

$$\frac{\partial c(0,t)}{\partial x} = \frac{\partial c(1,t)}{\partial x} = 0$$

<u>Problem</u>: Solve the set of PDEs with the given initial and boundary conditions. Plot the solution in two graphs $n(x,t)$ and $c(x,t)$.

First present the PDEs set in standard form:

$$\begin{bmatrix} 1 \\ 1 \end{bmatrix} \cdot * \frac{\partial}{\partial t}\begin{bmatrix} u_1 \\ u_2 \end{bmatrix} = \frac{\partial}{\partial x}\begin{bmatrix} \left(d\dfrac{\partial u_1}{\partial x} - a \cdot u_1 \dfrac{\partial u_2}{\partial x}\right) \\ \dfrac{\partial u_2}{\partial x} \end{bmatrix} + s \cdot \begin{bmatrix} r \cdot u_1 \cdot (N - u_1) \\ \dfrac{u_1}{u_1 + 1} - u_2 \end{bmatrix}$$

where u_1 and u_2 are n and c, respectively; as in the preceding example, the element-wise multiplication yields the vector $\begin{bmatrix} \dfrac{\partial u_1}{\partial t} \\ \dfrac{\partial u_2}{\partial t} \end{bmatrix}$.

Now present the initial and boundary conditions in standard form:

$$\begin{bmatrix} u_1(x,0) \\ u_2(x,0) \end{bmatrix} = \begin{bmatrix} \begin{cases} 1.05, & 0.3 \le x \le 0.6 \\ 1, & \text{otherwise} \end{cases} \\ \begin{cases} 0.50025, & 0.3 \le x \le 0.6 \\ 0.5, & \text{otherwise} \end{cases} \end{bmatrix}$$

$$\begin{bmatrix} 0 \\ 0 \end{bmatrix} \cdot * \begin{bmatrix} u_1 \\ u_2 \end{bmatrix}_a + \begin{bmatrix} 1 \\ 1 \end{bmatrix} \cdot * \begin{bmatrix} d\dfrac{\partial u_1}{\partial x} - a \cdot n\dfrac{\partial u_2}{\partial x} \\ \dfrac{\partial u_2}{\partial x} \end{bmatrix}_a = \begin{bmatrix} 0 \\ 0 \end{bmatrix}$$

$$\begin{bmatrix} 0 \\ 0 \end{bmatrix} \cdot * \begin{bmatrix} u_1 \\ u_2 \end{bmatrix}_b + \begin{bmatrix} 1 \\ 1 \end{bmatrix} \cdot * \begin{bmatrix} \dfrac{1}{2}\dfrac{\partial u_1}{\partial x} \\ \dfrac{1}{2}\dfrac{\partial u_2}{\partial x} \end{bmatrix}_b = \begin{bmatrix} 0 \\ 0 \end{bmatrix}$$

Published by Woodhead Publishing Limited

Here, as earlier, *a* and *b* are the boundary points.

To solve the problem, write the `tumour` function file:[4]

```
function tumour
%   To run:>> tumour
close all
m = 0;
xpoints = 81;
tpoints = 21;
x = linspace(0,1,xpoints); t = linspace(0,100,tpoints);
sol = pdepe(m,@tumour_pde,@tumour_ic,@tumour_bc,x,t);
n = sol(:,:,1); c = sol(:,:,2);
subplot(1,2,1)
surf(x,t,c);
title('Distribution of fibronectin, c(x)');
xlabel('Distance x'); ylabel('Time t');
axis square
subplot(1,2,2)
surf(x,t,n);
title('Distribution of ECs, n(x)');
xlabel('Distance x'); ylabel('Time t');
axis square
function [c,f,s] = tumour_pde(x,t,u,DuDx)
% PDEs for solution
d = 1e-3; a = 3.8;S = 3; r = 0.88; N = 1;
c = [1; 1];
f = [d*DuDx(1) - a*u(1)*DuDx(2);DuDx(2)];
s = [S*r*u(1)*(N - u(1)); S*(u(1)/(u(1) + 1) - u(2))];
function u0 = tumour_ic(x)
% Initial conditions
u0 = [1; 0.5];
if x > = 0.3 & x < = 0.6
  u0(1) = 1.05 * u0(1);
  u0(2) = 1.0005 * u0(2);
end
function [pa,qa,pb,qb] = tumour_bc(xl,ul,xr,ur,t)
% Boundary conditions
pa = [0; 0]; qa = [1; 1];
pb = [0; 0]; qb = [1; 1];
```

The `tumour` function is written without input and output arguments; within it, the `xmesh` and `tspan` vectors are created with two `linspace` commands for the selected numbers of x and t points: 81 and 21, respectively; $m = 0$ as described above. Again, three sub-functions are involved: the `tumour_PDE`, containing the commands for constants d, a, S, r and N; and two PDEs written in the two-row matrices c, f and s, namely the `tumour_ic`, defining the initial conditions in the two-row matrix u_0, and `tumour_bc` defining the conditions at $x = a$ and $x = b$. The numerical results are stored in the 3D `sol` array, rewritten as the two 2D matrices u and v for subsequent use in the graphical commands. Two `subplot` commands are used to place two graphs on the same page: the `surf` commands generate the 3D surface plots $u(x,t)$ and $v(x,t)$, and the `axis square` commands render the current axis box square in shape.

After running this function in the Command Window

>>tumour

the following plots are generated:

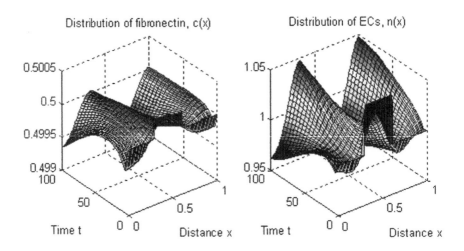

5.3 Questions for self-checking and exercises

1. Which ODE solver is recommended for an ODE, when the class of the problem is unknown in advance: (a) `ode23`, (b) `ode45`, (c) `ode15s`, (d) `ode113`?
2. For a stiff ODE, which of the following solvers should be tried first: (a) `ode15s`, (b) `ode15i`, (c) `ode23s`, (d) `ode23tb`?
3. The `tspan` vector specifying the interval of solution in the ODE solvers should have: (a) at least one value of the starting point of the interval, (b) at least two values – the starting and end points of the interval, (c) several values of the interval, including the starting and end

points, or are (d) the two latter answers correct, (e) the answers to (a) and (c) correct?

4. In the pdepe solver a vector xmesh specifies: (a) x and t coordinates to solve PDEs, (b) x coordinates of the points at which a solution is required for every time value, (c) the t coordinates of the points at which a solution is required for every x value?

5. How many and which functions should be written to solve a PDE with the pdepe solver: (a) one function with PDE, (b) two functions, with PDE and with the initial condition, (c) three functions, with PDE, with the initial condition and with the boundary condition?

6. Tumor growth rate is described by the following ODE:

$$\frac{dx}{dt} = ax - bx^2$$

where x is the size of the tumor. The coefficients a and b equal 0.9 month^{-1} and 0.7 month^{-1} cm^{-1}, respectively. Solve this equation numerically with the ODE solver in the time interval 0–8 months; initial value $x_0 = 0.05$ cm. Write a function with input and output arguments and save it in a function file. Display the x values obtained at $n = 8$ equidistant time points, and plot the x,t graph.

7. According to data for the 1918 influenza pandemic, the rate of spread of new infections is described by the equation

$$\frac{dR}{dt} = \frac{\mu}{\sigma^2}\left[R^{\left[\frac{\sigma}{\mu}\right]^2} - 1 \right]$$

where R is the number of new cases per day (called the basis reproductive number); the mean generation time $\mu = 2.92$ days and the variance $\sigma^2 = 5.57$ days^2 are statistical data. Solve this equation numerically with an ODE solver in the time interval 0–15 days with initial $R_0 = 20$. Write a function with input and output arguments and save it in a function file. Display the R values obtained at $n = 8$ equidistant time points, and plot the R,t graph.

8. Reproduction of the *Escherichia coli* bacterium on glucose is described by the following set of dimensionless ODEs:

$$\frac{d\bar{N}}{d\bar{t}} = \bar{N}\frac{\bar{S}}{1+\bar{S}}$$

$$\frac{d\bar{S}}{d\bar{t}} = -\bar{N}\frac{\bar{S}}{1+\bar{S}}$$

where $|\bar{S}| = S \cdot a$ is the dimensionless concentration of glucose, $|\bar{N}| = N \cdot a \cdot \gamma$ the bacteria population and $|\bar{t}| = r \cdot t$ is time; S and N are dimensional concentrations in g/L, and t is dimensional time in hours; the reaction rate constant $r = 1.35$ h^{-1}, and the constants $a = 0.004$ g/L and $\gamma = 0.23$.

Solve this set numerically in the time interval 0–6 h at steps of 0.5 h, take initial values $|\bar{N}|_0 = 0.25$ and $|\bar{S}|_0 = 2$. For solution use the ode23 function. Display the three-column array with dimensional t, S and N values. Plot the $\bar{S}(t)$ and $\bar{N}(t)$ curves.

9. In a batch reactor two reactions, $A{\to}P$ and $A+A{\to}U$, take place with rate constants $k_1 = 2$ for the first and $k_2 = 1$ for the second, in arbitrary units. P is the desired product and U an undesired product. These processes are described by the following set of ODEs:

$$\frac{d[A]}{dt} = -k_1[A] - k_2[A]^2$$

$$\frac{d[P]}{dt} = k_1[A]$$

$$\frac{d[U]}{dt} = k_2 A^2$$

where $[A]$, $[P]$ and $[U]$ are the amounts of species A, P and U, respectively.

Solve these equations numerically in the time interval 0–1 with initial $[A]_0 = 2$ and $[P]_0 = [U]_0 = 0$. Plot the $[A](t)$, $[P](t)$ and $[U](t)$ curves.

10. In pharmacokinetics the dissolution rates of a drug in the gastrointestinal (GI) tract and in the blood are described by the following set of two stiff equations:

$$\frac{dx}{dt} = -ax + D(t)$$

$$\frac{dy}{dt} = ax - by$$

where x and y are the amounts of drug in the GI tract and in the blood, respectively, a and b are the corresponding half-lives of the drug, and $D(t)$ is the dosing function. It is assumed that a drug is taken every 6 hours and dissolves within 30 min. $D(t)$ is described by the following equation

$$D = \begin{cases} 4, & t\text{-}6 \cdot \text{floor}(t/6) < 0.5 \\ 0, & \text{otherwise} \end{cases}$$

where D is in mg, and 'floor' denotes rounding towards minus infinity. The coefficients a and b are 1.38 h^{-1} and 0.138 h^{-1}, respectively.

Solve the set of ODEs numerically with zero initial drug amounts in the GI tract and in the blood. Time values are 0, 0.15, 0.3, 0.6, 0.9, 1, 1.5, 2, 2.5, 3, 4, 5 and 6 hours. Plot the $x(t)$ and $y(t)$ curves.

Published by Woodhead Publishing Limited

11. The classical predator–prey model describes the decline in proliferation due to prey crowding.[5] The set of equations for this model is

$$\frac{dx}{dt} = k_1 x - k_2 xy - k_5 x^2$$

$$\frac{dy}{dt} = -k_3 y + k_4 xy$$

where x and y are populations of prey and predators, the prey survival factor $k_1 = 1.0019$ year^{-1}, the prey death factor $k_2 = 1.00224$ year^{-1}, the predator survival factor $k_3 = 0.999914$ year^{-1}, the predator death factor $k_4 = 1.0058$ year^{-1} and the prey crowding factor $k_5 = 0.101331$ year^{-1}.

Solve these ODEs in the time interval 0–20 years with initial conditions $x_0 = 2000$ and $y_0 = 2000$. Plot the $x(t)$ and $y(t)$ curves, and x versus y in the phase plane graph.

12. An animal population growth rate is described by the differential equation

$$\frac{dN}{dt} = rN\left(1 - \frac{N}{K}\right)$$

where N is the animal population, t is time, K is the environmental carrying capacity and r is the growth rate coefficient. For the gray wolf population in one of the American states $K = 250$ animals and $r = 0.21$ year^{-1}.

Solve this equation numerically in the time interval 0–35 years with initial $N_0 = 15,100$ and 300 wolves. Plot all the results in one $N(t)$ graph, and use a constant number of the N-points for all $N(t)$ curves.

13. A bio-fuel (ethanol) is produced from plant tissue by hydrolyzing the cellulose and fermenting the resulting glucose. The process is described by the following equations:[6]

$$\frac{d[C]}{dt} = \frac{6.173 V_{max}[C]}{K_m\left(1 + 6.173\dfrac{[C]}{K_m} + 5.556\dfrac{[S]}{K_i}\right)}$$

$$\frac{d[X]}{dt} = [X]Ev,$$

$$\frac{d[S]}{dt} = -\frac{1}{y_{x/s}}\frac{d[X]}{dt} - 0.18\frac{d[C]}{dt}$$

$$\frac{d[P]}{dt} = v[X]$$

Published by Woodhead Publishing Limited

where $[C]$, $[S]$, $[X]$ and $[P]$ are the concentrations of the cellulose, substrate (glucose), enzyme cells and product (ethanol), respectively;

$$v = v_m \frac{[S]}{K_s + [S]}\left[1 - \frac{[P]}{[P]_{max}}\right]^n, \ v_m = E/\mu_{max} \ \text{and} \ V_{max} = k_p Et.$$

The constants in these equations are assumed: $Et = 1000$ mg protein, $k_p = 0.3$ mм glucose/(mg protein h), $K_m = 4$ mм, $K_i = 0.5$ mм, $K_S = 0.315$ g/L, $[P]_{max} = 87.5$ g/L, $\mu_{max} = 0.15$ g EtOH/(g cells h), $E = 0.249$, $Y_{X/S} = 0.07$ and $n = 0.36$.

Solve the set of ODEs in the time interval 0–10 hours with initial concentrations $[C]_0 = 125$ g/L, $[X]_0 = 1$ g/L and $[X]_0 = [P]_0 = 0$. Plot the $[C]$ and $[P]$ concentrations obtained versus time.

14. The diffusion in a biochemical reaction is described by the Brusselator-type PDEs:

$$\frac{\partial u}{\partial t} = 1 + u^2 v - 4u + \alpha \frac{\partial^2 u}{\partial t^2}$$

$$\frac{\partial v}{\partial t} = 3u - u^2 v + \alpha \frac{\partial^2 v}{\partial t^2}$$

where u and v are the species concentrations, and the diffusion coefficient $\alpha = 0.02$.

Solve this set of PDEs in the time interval 0–10 and coordinate interval 0–1 with the initial conditions

$$u_0 = 1 + \sin 2\pi x$$
$$v_0 = 3$$

and boundary conditions

$$\frac{\partial u(0,t)}{\partial x} = \frac{\partial u(1,t)}{\partial x} = 0$$

$$\frac{\partial v(0,t)}{\partial x} = \frac{\partial v(1,t)}{\partial x} = 0$$

Plot the u and v concentrations obtained as surfaces $u(t,x)$ and $v(t,x)$, taking the azimuth and elevation angles as –40° and 30°, respectively.

15. The concentration n of a bacterial culture which is diffusing (due to random motion) and reproducing, depending on the nutrient local concentration c, can be described by the following PDEs:

$$\frac{\partial n}{\partial t} = D_n \frac{\partial^2 n}{\partial t^2} + \left(\frac{k_{max} c}{k_n + c} - k\right) n$$

$$\frac{\partial c}{\partial t} = D_c \frac{\partial^2 c}{\partial t^2} - \alpha \frac{k_{max} c}{k_n + c} n$$

where D_n and D_c are diffusion coefficients, $\dfrac{k_{max}c}{k_n + c}$ is the Michaelis–Menten-type reproduction rate, k_{max} and k_n are constants, and k is the bacterial death rate coefficient. It is assumed that $D_n = 2$, $D_c = 1.2$, $k_{max} = 2$, $k_n = 1.9$, $\kappa = 0.8$ and $\alpha = 0.3$.

Solve these PDEs in the time interval 0–1 and the coordinate interval 0–1 with the initial conditions

$$n_0 = \begin{cases} 1, \text{ at } 0.4<x<0.6 \\ 0, \text{ otherwise} \end{cases}$$

$$c_0 = 2$$

and boundary conditions

$$\frac{\partial u(0,t)}{\partial x} = \frac{\partial u(1,t)}{\partial x} = 0$$

$$\frac{\partial v(0,t)}{\partial x} = \frac{\partial v(1,t)}{\partial x} = 0$$

Plot the surfaces of the $n(t,x)$ and $c(t,x)$ concentrations obtained on the same page, taking the azimuth and elevation angles as 130° and 45°, respectively.

5.4 Answers to selected exercises

3. (d) The two latter answers are correct.
5. (c) Three functions – with PDE, with initial conditions and with boundary conditions
8. `>> t_Glucose_Ecoli = Ch4_8(0.004,0.23,1.35,0,4.5,0.25,2)`
   t_Glucose_Ecoli =

0	0.0002	0.0080
0.3704	0.0003	0.0076
0.7407	0.0004	0.0071
1.1111	0.0006	0.0064
1.4815	0.0008	0.0055
1.8519	0.0011	0.0043
2.2222	0.0014	0.0031
2.5926	0.0016	0.0018
2.9630	0.0019	0.0009
3.3333	0.0020	0.0004
3.7037	0.0020	0.0001
4.0741	0.0021	0.0000
4.4444	0.0021	0.0000

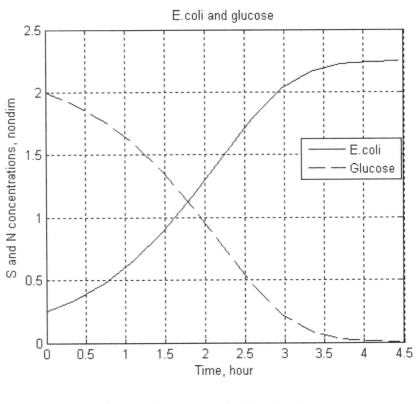

E.coli and glucose

10. >> t_GI_Blood = Ch4_10(1.38,0.138,0,6,0,0)
     t_GI_Blood =

0	0	0
0.1500	0.5420	0.0576
0.3000	0.9826	0.2143
0.4500	1.3409	0.4493
0.6000	1.2538	0.7190
0.7500	1.0191	0.9365
0.9000	0.8285	1.1058
1.0000	0.7219	1.1966
1.5000	0.3630	1.4621
2.0000	0.1822	1.5386
2.5000	0.0915	1.5233
3.0000	0.0460	1.4656
4.0000	0.0116	1.3083
5.0000	0.0029	1.1476
6.0000	0.0007	1.0017

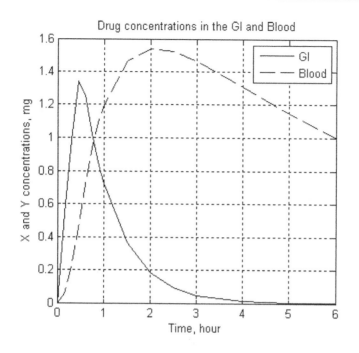

12. >> Ch4_12(0.21,250,0,35,[15 100 300])

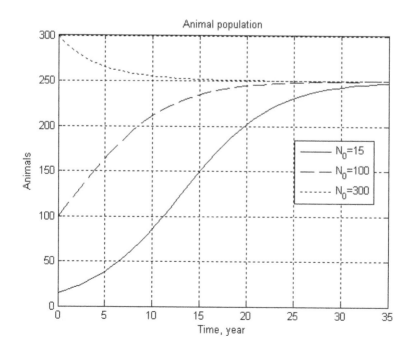

14. >> Ch4_14

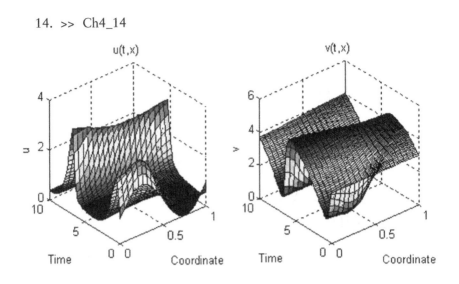

Notes

1. The options can be perused in greater detail by typing `help odeset` in the Command Window.
2. Also often referred to as a continuous stirred-tank bioreactor, CSTB.
3. Orme, M.E. and Chaplain, M.A.J. (1996) A mathematical model of the first steps of tumor-related angiogenesis: capillary sprout formation and secondary branching. *IMA Journal of Mathematics Applied in Medicine & Biology* 13: 73–98.
4. The `tumour` function follows the `pdex5` example given in MATLAB® demo; to view it, type and enter `>>edit pdex5`.
5. Compare this model with that discussed in subsection 5.1.3.3
6. The problem is reproduced from Mosier, N.S. and Ladish, M.R. (2009) *Modern Biotechnology. Connecting Innovations in Microbiology and Biochemistry to Engineering Fundamentals.* New York: Wiley, p. 136.

6

Bioinformatics tool for sequence analysis

In present-day genetics, molecular biology, biomedicine and other areas of the life sciences, computers are used for studying biotic systems and bioprocesses. The discipline involved, bioinformatics, operates through accessing and analysis of information accumulated in formatted databases. For this, MATLAB® offers a specially designed tool – the Bioinformatics toolbox™ containing a set of specialized commands, some of which are discussed in this chapter.

In this context, terms and information relevant to biology, database organization and sequence analysis are given here briefly. For further details, biology and bioinformatics tutorials should be consulted.

6.1 About toolboxes

Although commands and functions such as `sin`, `cos`, `sqrt`, `fzero`, `quard`, `save`, `ode45` and `pdpde` are operative in a wide range of areas from mechanics to medicine, specialized means are needed in each area for its specific problems. Basic and problem-oriented tools are collected in so-called toolboxes intended for particular engineering areas; for example, basic commands are assembled in the MATLAB® toolbox, commands related to signal processing in the Signal Processing toolbox and commands for neural networks in the Neural Network toolbox. To find out which toolboxes are available in your computer, the `ver` command should be used. Typing this in the Command Window and then entering it, the mathworks product family header information will display together with a list of toolbox names, versions and release. The list can be very long, depending on the MATLAB® configuration on the actual computer:

```
>> ver
-----------------------------------------------------------------------
MATLAB® Version 7.10.0.499 (R2010a)
MATLAB® License Number: 617196
Operating System: Microsoft Windows XP Version 5.1 (Build
2600: Service Pack 3)
Java VM Version: Java 1.6.0_12-b04 with Sun Microsystems
Inc. Java HotSpot(TM) Client VM mixed mode
-----------------------------------------------------------------------
MATLAB®                     Version 7.10      (R2010a)
Simulink                    Version 7.5       (R2010a)
Aerospace Blockset          Version 3.5       (R2010a)
Aerospace Toolbox           Version 2.5       (R2010a)
Bioinformatics Toolbox      Version 3.5       (R2010a)
...
```

The information about toolboxes that is available can also be obtained from the pop-up menu appearing after clicking the Start button in the bottom line of the MATLAB® Desktop – see Figure 2.2.

Initially, bioinformatics was confined to genomics and genetics, in particular DNA sequences. Today, it covers compilation and advancement of databases, algorithms, and computational and statistical techniques.

6.2 The functions of the Bioinformatics toolbox™

The toolbox performs the following functions:

- access to public databases on the Web and other online data sources;
- support formats specific to genomic, proteomic and gene expression data;
- transfer of data, written in different bioinformatics formats, directly to the Workspace for further processing by MATLAB® programs;
- analysis of DNA, RNA and peptide sequences. A sequence represents a frequently very long chain of letters chosen from the four-letter DNA and RNA or 23-letter amino-acid alphabet; for more details see subsection 6.4.1.
- analysis of microarray or phylogenetic data;
- preprocessing of mass spectrometry data;

- reading data generated by gene sequencing instruments, mass spectrometers and microarray scanners;
- visualization of sequences, sequence statistics, etc.

6.3   Public databases, data formats and commands for their management

Biological information is collected in special formats and can be retrieved from databases that support data exchange and processing. Web-based databases are open to the public and can be copied to and processed on your computer.

6.3.1   Databases to which MATLAB® has access

Currently, the toolbox provides access to the GenBank, GenPept, EMBL, PDB, NCBI GEO and PFAM databases.

GenBank stores an annotated collection of the genetic sequences – it is managed by the National Institute of Health (USA); GenPept contains translated protein-coding sequences – it is managed by the National Center for Biotechnology Information (NCBI), which provides this Entrez system for protein information search; EMBL belongs to the European Molecular Biology Laboratory and stores Europe's primary nucleotide sequence resources; PDB (Protein Data Bank) stores 3D structural data on large biological molecules, namely proteins and nucleic acids – it is managed by the Worldwide PDB organization; GEO (Gene Expression Omnibus) stores chips, microarrays, gene expression data and hybridization arrays – it is supported by the National Center for Biotechnology Information, NCBI; the GO (Gene Ontology) database stores gene product properties, including the PFAM database that contains information about protein families with their annotations and multiple sequence alignments generated by the hidden Markov model – it is managed by the Wellcome Trust Sanger Institute.

6.3.2   About formats for storage and searching of database information

For each stored substance the data set contains sequences, text, graphs, numbers and other information written in different data formats. For operations with such diverse data MATLAB® uses so-called structures, which are commonly taught in advanced courses; only the minimal information about structures required to use the toolbox commands is provided here.

A structure is constructed from the name of the variable and those of the fields, which are separated from the structure name by a period; for example, s.sequence is the structure named s with the sequence field, and s.molar_weight is the molar_weight field of the same s structure. The fields contain data of different types; for example, in the s structure the sequence field may contain a set of characters of the char type, and the molar_weight may contain a molar-weight value of the double (numerical) type. The number of fields is limited only by the amount of memory. They can contain tables, plots or subfields with data. The structure may be given as a vector, in which case for reading or writing the data, the index of the element should be provided. For example, s(2).sequence means sequence data from the second element of the s structure.

The data are stored in databases in particular form defined by a special file format. Below are some file formats supported by the Bioinformatics toolbox™ for sequence data:

- FASTA is intended for coding nucleotide and peptide sequences represented by single-letter codes;
- PDB is a file format designed for storage of protein information with the 3D structure of protein and nucleic acid molecules;
- SCF (source comparison file) is a format of data transmitted from instruments used for DNA sequencing.

The toolbox also supports additional file formats such as: the Affimetrix DAT, EXP, CEL, CHP, GenePix GRP and GAL for microarray data; Microsoft Excel or CSV for industrial-specific data; and various others for web-based databases. More detailed information is available in the original toolbox documentation.

Working with databases necessitates recourse to specialized algorithms searching for similarity between sequences; the most widespread are FASTA (fast algorithm; the same name as the format described above) and BLAST (Basic Local Alignment Search Tool). FASTA is intended for alignment of DNA and protein sequences, and BLAST for comparing nucleotide or protein sequences with sequence databases and calculating the statistical significance of matches.

6.3.3 Commands for accessing and reading database files

For each format of data stored in a database, MATLAB® offers a set of commands for receiving the data, saving it to a file in the specified format, and reading from the file all previously saved formatted data. For retrieval of information, each biological species has to be assigned an identifier to be obtained from the bank before the request is placed.

For example, nucleotide information may be retrieved from the CenBank© and written into the variable or into the MATLAB® file by following the simplest forms of the `getgenbank` command:

```
data = getgenbank('accession_number')
getgenbank('accession_number','ToFile','file_name')
```

where `accession_number` is the identifier of a sequence accepted in the bank; `ToFile` is the property name to be written into the quotes; `file_name` is a string with file name for saving GenBank data in the current directory, or with the full pass and file name for saving in any other directory; `data` is a structure that contains information about a sequence received from the bank.

The accession number (called locus number in some databases) is an important attribute that should be defined before transmitting information from a database to the MATLAB® Workspace.

For database exploration and to obtain an accession number, the `web` command can be used. Its simplest form is

```
web('url')
```

where `url` (uniform resource locator) is a string defining the Internet address of the searched resource. The `web` command opens the MATLAB® Web browser with the website indicated by `url`; for example, if we want to open NCBI databases, we should type and enter in the Command Window:
>>web('http://www.ncbi.nlm.nih.gov/')

The browser window with NCBI home page (Figure 6.1, p. 176) is opened:
The window provides two Search boxes. For instance, if we need an accession number for stored information about the hexosaminidase A (HEXA) enzyme for the Norway rat, we select 'Nucleotide' in the first box pop-up menu (appears after pressing the arrow on the right of the

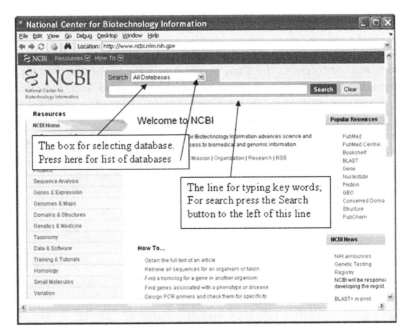

Figure 6.1   Matlab Web Browser with NCBI home page.

box – see Figure 6.1) and the key words 'hexosaminidase A Norway rat' should be typed in the second Search line. The window that is opened is shown in Figure 6.2 (p. 177).

This accession number should be used in the `getgenbank` command that retrieves sequence information about the Norway rat and transmits it to the S structure:

```
>> % transmits from database to the S-structure
>> % sequence information about the Norway rat
>> S=getgenbank('NM_001004443')

S =
                  LocusName: 'NM_001004443'
        LocusSequenceLength: '1779'
       LocusNumberofStrands: ''
              LocusTopology: 'linear'
          LocusMoleculeType: 'mRNA'
       LocusGenBankDivision: 'ROD'
     LocusModificationDate: '26-JAN-2010'
```

Published by Woodhead Publishing Limited

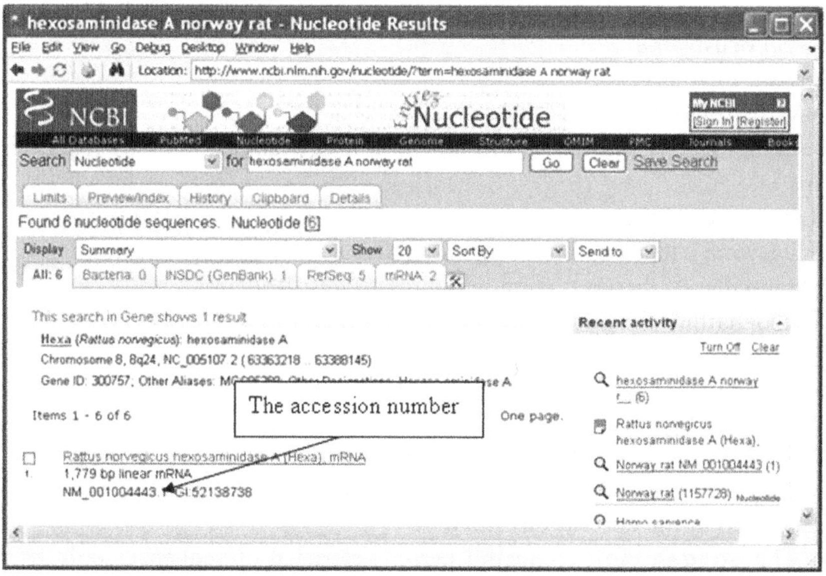

Figure 6.2   The page with accession number for rat hexosaminidase A.

```
       Definition: [1x48 char]
        Accession: 'NM_001004443 XM_217144'
          Version: 'NM_001004443.1'
               GI: '52138738'
          Project: []
           DBLink: []
         Keywords: []
          Segment: []
           Source: 'Rattus norvegicus (Norwayrat'
   SourceOrganism: [4x65 char]
        Reference: {1x3 cell}
          Comment: [9x66 char]
         Features: [48x74 char]
              CDS: [1x1 struct]
         Sequence: [1x1779 char]
        SearchURL: [1x73 char]
      RetrieveURL: [1x103 char]
```

To display information contained in one of the fields of the S structure –
e.g. in the LocusName and in the SourceOrganism fields – type and enter:

\>> display an access number that contains in the LocusName field
\>> S.LocusName
ans =
NM_001004443
\>> S. SourceOrganism   % display the source organism for this sequence
ans =
Rattus norvegicus
Eukaryota; Metazoa; Chordata; Craniata; Vertebrata; Euteleostomi;
Mammalia; Eutheria; Euarchontoglires; Glires; Rodentia;
Sciurognathi; Muroidea; Muridae; Murinae; Rattus.

\>> getgenbank('NM_001004443','ToFile', 'Norway_Rat.txt')

After entering the latter command the same information as in the previous
getgenbank example is displayed and saved in the Norway_Rat.txt file
generated in the current directory.

The getgenbank command permits retrieval of sequences only from
the database:

```
data=getgenbank('Accession_Number','SequenceOnly','true')
```

For reading files saved as ASCII text files in the GenBank format, the
genbankread command is used:

```
S_data=genbankread('file_name')
```

where S_data is a structure with fields corresponding to the GenBank
keywords; and file_name is a string with a file name.

An example of this command form usage is:
\>> S_data=genbankread ('Norway_Rat.txt')   % see precede example
S_data =
1x2 struct array with fields:
   LocusName
   LocusSequenceLength
   LocusNumberofStrands
   LocusTopology
   LocusMoleculeType
   LocusGenGankDivision
   LocusModificationDate
   Definition
   Accession
   Version

GI
Project
DBLink
Keywords
Segment
Source
SourceOrganism
Reference
Comment
Features
CDS
Sequence
>> displays accession number contained in the LocusName field
>>S_data.LocusName
ans =
NM_001004443

Some additional commands for database management are shown in Table 6.1.

Table 6.1 Additional commands for database management*

Form of MATLAB presentation	MATLAB example (inputs and outputs)

data=
getgenpept('Accession_Number')   or
`getgenpept('Accession_`
`Number', 'ToFile','File_Name')`
gets from GenPept database sequence data for a substance with given Accession_Number and writes it in the data structure or to the File_Name file; File_Name – is a string
	>>data=getgenpept('NP_663358')%
	domestic mouse
	data =
	LocusName: 'NP_663358'
	LocusSequenceLength: '348'
	LocusNumberofStrands: ''
	LocusTopology: 'linear'
	LocusMoleculeType: ''
	LocusGenBankDivision: 'ROD'
	LocusModificationDate: '28-FEB-2010'
	
	... **

data
`=getembl('Accession_Number')`
or
`getembl('Accession_Number',`
`'tofile','File_Name')`
gets sequence information from EMBL database;
the arguments here are the same as for the `getgenpept` command
	data=getembl('AJ007347')% Foot/
	Mouse Virus
	data =
	Identification: [1x1 struct]
	Accession: 'AJ007347'
	SequenceVersion: 'AJ007347.1'
	DateCreated: '11-DEC- 1998
	(Rel. 58, Created)'
	... **

Table 6.1 *Continued*

`data = getpdb('identifier')`
 or
 `getpdb(identifier,'ToFile',`
 `'File_Name')`
 gets protein 3D structure information from the PDB database. The `identifier` in this database consists of four characters, e.g. 3E3E is the identifier for human thioredoxin double mutant. The other arguments are the same as for the `getgenpept` command
	`>>data=getpdb('3E3E','ToFile', 'human_`
	 `mutant.pdb')`
	 Data =
	 Header: [1x1 struct]
	 Title: [1x41 char]
	 Compound: [7x57 char]
	 Source: [10x39 char]
	 Keywords: [2x59 char]
	 Experimentdata: 'X-RAY DIFFRACTION'
	 Authors: 'G.HALL,J.EMSLEY'
	 … **

`data=`
 `Getgeodata('Accession_`
 `Number')`
 or
 `getgeodata('Accession_`
 `Number', 'ToFile', 'File_Name')`
 	
 gets information from the GEO database.
 The arguments are the same as for the `getgenpept` command
	`>>% studied organism – home mouse`
	 `>>%what studied: liver response to`
	 human and `>>%chimpanzee diets`
	 `>> data =getgeodata('GDS3221')`
	 data =
	 Scope: 'DATASET'
	 Accession: 'GDS3221'
	 Header: [1x1 struct]
	 ColumnDescriptions: {24x1 cell}
	 ColumnNames: {24x1 cell}
	 IdRef: {45101x1 cell}
	 Identifier: {45101x1 cell}
	 Data: [45101x24 double]

`S_data=`
 `genpeptread('File_Name.txt')`
 assigns data from the GenPept-formatted `File_Name.txt` file to the `S_data` structure
	`>> %Human growth hormone`
	 `>> getgenpept('1HGU_A',`
	 `'ToFile','human_hormone.txt');`
	 `>> S_data=genpeptread('human_`
	 `hormone.txt')`
	 S_data =
	 LocusName: '1HGU_A'
	 LocusSequenceLength: '191'
	 LocusNumberofStrands: ''
	 LocusTopology: 'linear'
	 LocusMoleculeType: ''
	 LocusGenBankDivision: 'PRI'
	 LocusModificationDate: '10-JUL-2009'
	 Definition: 'Chain A, Human Growth Hormone.'
	 … **

Table 6.1 *Continued*

`S_data`
`=emblread('File_Name.txt')`
assigns data from the EMBL-formatted
`File_Name.txt` file to the `S_data-` structure.
	`>>%Human Protein`
	`>>getembl('X79439','ToFile','human_protein.txt');`
	`>> S_data=emblread('human_protein.txt')`
	`S_data =`
	　　　Identification: [1x1 struct]
	　　　Accession: 'X79439'
	　　SequenceVersion: 'X79439.1'
	　　　DateCreated: '05-APR-
	　　　　　　1995 (Rel. 43,
	　　　　　　Created)'
	`...**`

`S_data=`
`Pdbread('File_Name.pdb')`
assigns data from the PDB-formatted
File_Name.pdb file to the S-data-structure.
	`>> % human_mutant.pdb file created by the getpdb`
	`>> dataa=pdbread('human_mutant.pdb')`
	`Dataa =`
	　　　Header: [1x1 struct]
	　　　Title: 'HUMAN
	　　　　　THIOREDOXIN
	　　　　　DOUBLE MUTANT
	　　　　　C35S,C73R'
	　　Compound: [7x57 char]
	　　　Source: [10x39 char]
	`...**`

`molviewer('human_mutant.`
`pdb')`
displays and manipulates 3D molecular structures.
	`>> % human_mutant.pdb file created by the getpdb`
	`>>molviewer('human_mutant.pdb')`

`RID=blastncbi('Seq',`
`'Program')`
requests report on the `Seq` -nucleotide or amino acid sequence. Report is created in the NCBI BLAST environment by the `Program`, that can be 'blastp' (for protein), 'blastn' (for nucleotide), and others, obtainable with the help.
	`>> %Rat insulin receptor`
	`>>RID = blastncbi('AAA41441','blastp')`
	`RID =`
	`VNZJP7K3013`

Table 6.1 *Continued*

The 'exact' property can be introduced in this command next to the Seq argument, recommended exact values are 10^{-7}... 10^{-10}. Identification number of the BLAST report is assigned to the RID variable.	
S_data = getblast(RID) or getblast(RID, 'ToFile','file_name') gets a report with the RID identification number from the database to the File_Name text file; File_Name – is a string; S_data is a structure with report data. Note: complete information about the S_data fields should be obtained by the doc command	%RID obtained by preceding blastncb >> S_data = getblast(RID) S_data = RID: 'VNZJP7K3013' Algorithm: 'BLASTP 2.2.23+' Query: [1x385 char] Database: [1x114 char] Hits: [1x50 struct] Statistics: [1x1051 char]

*This table does not contain all possible command forms; for additional information use the help command.
**Shortened.

Each bank access command permits retrieval of a sequence only, which should be done exactly as explained above for the getgenbank command.

6.4 Sequence analysis

Computerized searching and study of nucleotide or amino acid sequence data is termed sequence analysis, which is used for gene identification, detection of inter-gene similarity and gene functions, and search for gene-coded proteins. These are realized with commands that fall under three groups – sequence utilities, sequence statistics and sequence alignment.

6.4.1 Sequence utilities and statistics

Utilities

This group of commands serves for manipulation of sequences for more effective use, deeper insight into their characteristics and statistical information.

One command of this group is randseq, frequently used in tests or for explanation of other statistical and analysis commands. It has the form:

```
seq_name=randseq(n, 'alphabet','alphabet_name')
```

where `seq_name` is the name of a variable containing the generated string of the letter-code of nucleotides or amino acids; n is a number that specifies the length of the random sequence; `'alphabet'` is the property specified by the `'alphabet_name'` value that can be written as `'dna'`, `'rna'`, or `'amino'` (can also be written as `'aa'`). The `'dna'` alphabet uses the letters A, C, G and T, the `'rna'` uses the letters A, C, G and U, the `'amino'` alphabet uses the letters A, R, N, D, C, Q, E, G, H, I, L, K, M, F, P, S, T, W, Y, V, B, Z and X, and the '*' and '-' symbols.

The `'dna'` alphabet is a default value; in this case the `'alphabet'`... `'alphabet_name'` option can be dispensed with.

For example, to generate a protein sequence of 30 letters the following command should be typed and entered in the Command Window:

\>> Protein_Seq=randseq(30,'alphabet','amino')

Protein_Seq =

TYNYMRQLVVDVVITNHYSVFATYFSPGFD

Other sequence manipulation commands are those for conversion to sequences using the genetic code. One of these commands is `aa2nt`, which converts an amino acid sequence to a nucleotide sequence; its simplified form is:

```
nucleo_seq= aa2nt(amino_seq,'GeneticCode',code_
          number,...'alphabet','alphabet_name')
```

where `nucleo_seq` is the nucleotide sequence represented by a character string; `amino_seq` is the string or numeric code specifying the amino acid sequence or the structure with a `sequence` field containing the amino acid sequence previously retrieved by one of the access or read commands (as described in the preceding subsection); `'GeneticCode'` is a property specified by the `code_number` value; and `'alphabet'` is a property specified by the nucleotide `'alphabet_name'` string that should be typed as `'dna'` or `'rna'`. The `code_number` of the `GeneticCode` may range from 1 (default value) to 23 and should be chosen from a table obtainable by entering in the Command Window the `help aa2nt` command. If the `'dna'` alphabet is used, the `'alphabet'` option may be dispensed with.

Examples are:

\>> nucl_seq1=aa2nt('LifeSciense','alphabet','rna')

nucl_seq1 =

CUCAUUUUUGAGAGUUGCAUUGAGAAUUCUGAA

>>nucl_seq2=aa2nt('LifeSciense') % default: 'dna' alphabet and 1 genetic code
nucl_seq2 =
TTAATATTTGAGTCTTGCATTGAGAACAGTGAA

If the converted sequence contains one of '*', '-' or '?', a warning appears:
>>% 2 codes the Vertebrate Mitochondrial
>>nucl_seq=aa2nt('BI-SCIENCE','GENETICCODE',2,'Alphabet','rna')
Warning: The sequence contains ambiguous characters.
> In aa2nt at 122
nucl_seq =
GAUAUU---AGUUGCAUUGAGAACUGUGAA

To perform sequence comparison it is often convenient to transform the nucleotide sequence into a protein sequence, thereby reducing the 64 codons to 20 distinct amino acids. This can be done via the nt2aa command, which has the same form as aa2nt; for example, prot_ seq=nt2aa(nucl_seq2) converts the nucl_seq2 nucleotide sequence from one of the preceding examples to the prot_seq protein sequence.

In a gene sequence, non-coding sections may be mixed with exons and determination of a protein-coding sequence can be difficult. To identify the start and stop codons of the protein-coding sequence section (called open reading frame – ORF), you should read the sequence into the workspace with the seqshoworfs command, which defines their positions, assigns the values to a structure with the start and stop fields, and displays the Open Reading Frames window.

Statistical commands

The commands of this group are intended for statistical problems: nucleotide, codon, dimer and amino-acid counts in sequences, molar weight determination, calculation and plotting of nucleotide density changes along the sequence, graphical representation, etc.

As an example, we consider the aacount command that counts, displays and plots the amount of each amino acid in the sequence. Its simplified form is:

```
amino_s=aacount(amino_seq,'chart',chart_value)
```

where the amino_seq is a letter string or a row vector of integer numbers representing an amino acid sequence, or a structure with a sequence field that contains such a sequence previously constructed by the appropriate command described in Subsection 6.3.3; the 'chart' property is specified by the chart_value, and can be 'pie' or 'bar' depending on the desired graphical representation; the 'chart'-chart_value option can be dispensed with – in this case the graph is not generated; the amino_s is a structure containing 20 fields for each of the standard amino acids.

Below are versions of the aacount command:

```
>> % generation of the 25 letter sequence
>> seq=randseq(25,'alphabet','amino')
>> amino_acid=aacount(seq)
seq =
RCYDTLVRHNVASTWRGQTHYDQNN
amino_acid =
    A: 1
    R: 3
    N: 3
    D: 2
    C: 1
    Q: 2
    E: 0
    G: 1
    H: 2
    I: 0
    L: 1
    K: 0
    M: 0
    F: 0
    P: 0
    S: 1
    T: 3
    W: 1
    Y: 2
    V: 2
>> returns amount of Tryptophan (W) in the seq sequence
>> amino_acid.W
ans =
    1
>> assigns results to amino-acid and plots
>> amino_acid = ...
aacount(seq, 'chart', 'bar');
```

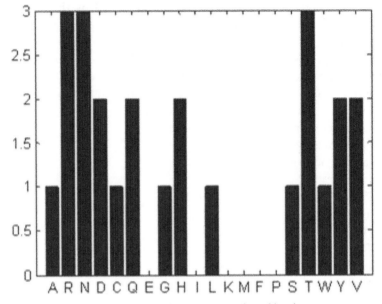

Figure 6.3   Bar chart of amino acid amounts produced by the `aacount` command for a randomly generated 25-letter sequence.

The first command of the example generates a random sequence of the amino acid letter codes and assigns the result to the `seq` variable. The second `aacount` command counts the amount of each amino acid in the sequence, assigns it to the appropriate fields of the `amino_acid` structure and generates the plot shown in Figure 6.3.

Another command of the statistical group is the `ntdensity`, which counts and plots the density of the A, C, G and T nucleotides. Its simplest form is:

```
nucl_dens=ntdensity(nucleotide_seq)
```

where `nucl_dens` is a structure with the fields A, C, G and T containing their densities; the `nucleotide_seq` is a letter string or a row vector of integer numbers representing a nucleotide sequence, or a structure with a `sequence` field that contains nucleotide sequence previously constructed by an appropriate command described in Subsection 6.3.3.

For example:
```
>> % get data for rhesus macaque
>> macaque=getgenbank('NM_001168654','sequenceonly','true');
 >> nucl_dens =ntdensity(macaque)
nucl_dens =
   A: [1x2345 double]
```

C: [1x2345 double]
G: [1x2345 double]
T: [1x2345 double]

The getgenbank command in this example retrieves the sequence for rhesus macaque semenogelin from the GenBank database and assigns the results to the macaque variable; the ntdensity command counts the A, C, G and T nucleotide density changes along the sequence, assigns them to the appropriate fields of the nucl_dens structure and generates the plots in Figure 6.4.

Some additional utility and statistical commands are presented in Table 6.2.

6.4.2 Sequence alignment

The sequence alignment technique is used to search for functional, structural or evolutionary similarity between sequences. As sequences are represented by a letter-coded row, the alignment consists of letter-by-letter comparison of two or more sequences.

The Bioinformatics toolbox provides a number of possibilities for alignment of paired and multiple sequences.

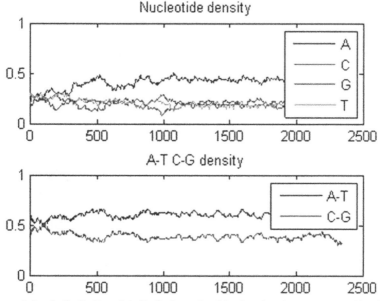

Figure 6.4 A, C, G, T and A, T, C, G nucleotide density plots generated by the ntdensity command for the rhesus macaque semen sequence.

Table 6.2 Additional utility and statistical commands*

Form of MATLAB presentation	MATLAB Example (inputs and outputs)
`aminolookup(seq)` – converts three-letter code abbreviations of the amino acids in the sequence to one-letter codes Notes: 1. `Aminolookup` has additional arguments for conversion of amino acid sequences written in one-letter-, three-letter, number- and codon codes; for more information type and enter `help aminolookup` in the Command Window. 2. `aminolookup` (written without arguments) displays a table of amino acids and their one-letter-, three-letter, number- and codon codes	>> seq=randseq(16,'alphabet','amino') seq = EYGNSGCHRNYVKRCG >> aminolookup(seq) ans = GluTyrGlyAsnSerGlyCys HisArgAsnTyrV alLysArgCysGly
`baselookup('Code', Code_Value)` displays nucleotide codes; Note: 1. The command has other arguments for complement sequence or integer codes 2. `baselookup` (written without arguments) displays a table of the nucleotide letter and integer codes, names, and complements	>> baselookup('code','T') ans = T Thymine >> baselookup('code','R') Ans = purIne G|A

`base_s= Basecount('nucl_seq','chart', 'chart_name')` counts the nucleotides in the `nucl_seq` sequence, assigns the result to the base_s structure with the A, C, G and T fields; generates a plot with `chart_name` equal to `'pie'` or `'bar'`.	>>
	macaque=getgenbank('NM_001168654', 'sequenceonly','true') >>
	>>base_macaque=basecount(macaque,'chart','pie')
	base_macaque =
	  A: 947
	  C: 484
	  G: 436
	  T: 478

Table 6.2 *Continued*

Form of MATLAB presentation	MATLAB Example (inputs and outputs)
`dimer_s=` `dimercount(nucl_` `seq,'chart','chart_name')` counts the dimers in the `nucl_seq` sequence, assigns the result to the dimer_s structure containing the AA, AC, AG, AT, CA, CC, CG, CT, GA, GC, GG, GT, TA, TC, TG and TT fields; generates a plot with chart_name equals `'pie'` or `'bar'`.	>> dimer_macaque=dimercount(macaque,' chart','bar') dimer_macaque = AA: 391 AC: 145 AG: 217 AT: 194 CA: 259 CC: 103 CG: 21 CT: 101 GA: 162 GC: 85 GG: 103 GT: 85 TA: 135 TC: 150 TG: 95 TT: 98 Note: the macaque sequence is taken from the basecount example
`codon_s=codoncount(nucl_` `seq,'figure','true')` counts the codons in the nucl_seq sequence, assigns the result to the codon_s structure containing 64 fields with possible codons (AAA, AAC, AAG, ...); generates a plot of the codon count in which the codon amounts are given by the color ruler on the right of this plot	>>codon_macaque=codoncount (macaque,'figure','true') Codon_macaque = AAA: 62 AAC: 28 AAG: 63 AAT: 16 ACA: 14 ACC: 18 ACG: 3 ...**
`molweight(amino_seq)` calculates the molecular weight of the `amino_seq` sequence	>> wild_boar =getgenbank('NM_001003662', 'sequenceonly','true') >>% follistatin (FST) >> molweight(wild_boar) ans = 8.4799e+004

*This table does not contain all possible command forms; for additional information use the `help` command.
**Shortened.

Pairwise alignment

In the process of alignment, gaps can be inserted into the sequences and characters can be inserted and/or deleted. The degree of identity of the sequences is measured by an alignment score criterion. A DNA example is shown in Figure 6.5, with matched characters (vertical bars) graded 1, mismatched characters and inserted gaps by −1.[1] The maximal score means the best alignment.

When two sequences are compared, a matrix is formed in which each row represents a character from one sequence and each column a character from the other; the numbers at a row–column intersection are the similarity score of the row and column character related to this intersection. The path through the maximal score values is the winning path, representing the best alignment. There are PAM, BLOSUM and other widely used scoring matrices for sequence alignment. In multiple alignments distance scores are used; in this case each pairwise score represents the grade of non-identical characters in the compared sequences.

Comparison of sequence alignments achieved via the various scoring systems can be improved by normalizing the scored values and the statistical properties of the system, which makes for non-integer score values. For evaluation of score reliability two statistical indices are often used – probability P and expected value E. A high level of sequence similarity is associated with a high P value (e.g. 90%) and a low E value (e.g. 10^{-7}), indicating that it is unlikely for the alignment to be randomly successful. The example in Figure 6.5 involves very short sequences, which can be

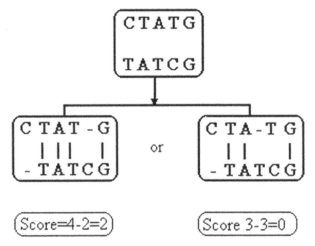

Figure 6.5   Sequence score example.

processed manually; real-life biological sequences are significantly longer and require specialized algorithms, classified as global and local. In the first case the goal is total sequence similarity and in the second the goal is to identify regions of similarity within the sequences.

For global alignment the Needleman–Wunsch algorithm is used; this performs by the `nwalign` command, whose simplest form is

```
[score,alignment]=nwalign(seq_1,seq_2,'showscore',true)
```

where `seq_1` and `seq_2` are the names of variables containing two amino acids or nucleotide sequences; the `'showscore'` property with argument `true` (written without quotes) opens the Figure window displaying the scoring space and the winning path of the alignment. The color of each cell in the scoring space map represents the best score determined over all possible alignments. The winning path is represented by black dots and is shown as the best position of the letter pairs; the color of the lower right cell of the winning path represents the optimal global alignment score, whose numerical value is transmitted to the `score` variable; `score` is optimal global score in bits; the `alignment` is a three-row array which displays in the first and third rows the `seq_1` and `seq_2` sequences, respectively, and in the second row the 'l', ':' or space symbols signing exactly matched (appear in red), mismatched (accepted as related by the scoring matrix, appear in magenta) and non-matched (zero or positive score, appear in black) letters.

In the presented form of `nwalign`, the BLOSSOM50 (default) scoring matrix is used for amino acids. Its expanded form can use other scoring matrices and some additional scoring and sequence characteristics; for detailed information, type the `help nwalign` command in the Command Window.

The command can be used without the `'showscore'` property, in which case only the score value and three rows of the `alignment` matrix are displayed in the Command Window.

An example for this command is:
```
>> sequence should contain the letters of the 'amino' alphabet
>> s_1='lifescience';
>> s_2='sefesici';        % the same requirement as for the s_1
>> [score, alignment]=nwalign(s_1, s_2, 'showscore',true)
score =
    0.6667
alignment =
```

```
LIFESCIENCE
 ||| |  |
SEFES-I-CI
```

After entering the latter command the Figure window appears (Figure 6.6) with the scoring space and winning path:

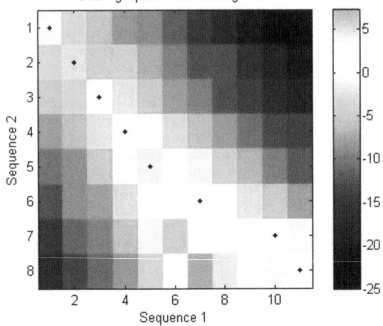

Figure 6.6 Scoring space heat map (colored cells and bar, to the right) and winning path (closed circles) generated with the `nwalign` command.

To align nucleotide sequences with this command the `'alphabet'` property with the `'nt'` argument should be added; for example, the last command in the preceding example can be rewritten as

```
[score, alignment]=nwalign(s_1, s_2,
        'showscore',true,'alphabet','nt')
```

where s_1 and s_2 are assumed here to be nucleotide sequences.

For local alignment the Smith–Waterman algorithm is used. This is realized in the `swalign` command and is similar to the previous command:

```
[score,alignment]=swalign(seq_1,seq_2,'showscore',true)
```

All input and output parameters here are the same as for the `nwalign` command with the only difference that the output values are the results of the local not the global alignment.

An example of this command application is:

```
>> % s1,s2 are the same as in the nwalign
>> [score, alignment]=swalign(s_1,s_2,'showscore',true)
score =
    9.6667
alignment =
FES-CI
||| ||
FESICI
```

The resultant map (Figure 6.7) is:

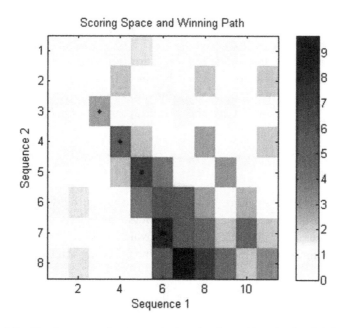

Figure 6.7   Scoring space heat map (colored cells and bar, to the right) and winning path (closed circles) generated with the `swalign` command.

Published by Woodhead Publishing Limited

Another graphical representation, often used for comparison of protein sequences, is the dot plot, sometimes called the residue contact map. The compared protein sequences are located along the vertical and horizontal axes, and identical proteins are shaded in black. If the plot clearly shows a diagonal line, the alignment is good. For generation of a dot plot the `seqdotplot` command is used:

$$\text{seqdotplot(seq\_1,seq\_2,window,number)}$$

where `seq_1` and `seq_2` are nucleotide or amino acid strings or structures with the `sequence` field obtained from a sequence database; `window` and `number` are integer numbers that define the size of a window and the number of characters that should be matched within it. For nucleotide sequences `window` is often taken as 11 and `number` as 7.

An example is:

```
>> mouseHEXA=getgenbank('AK080777','sequenceonly',true);
>> ratHEXA=getgenbank('NM_001004443','sequenceonly',true);
>> seqdotplot(mouseHEXA,rathexa,11,7)
Warning: Match matrix has more points than available screen pixels.
        Scaling image by factors of 2 in X and 2 in Y
>> xlabel('Mouse hexosaminidase A');
>> ylabel('Rat hexosaminidase A');
```

In this example the nucleotide sequences for domestic mouse and Norway rat hexosaminidase A are obtained by the `getgenbank` command. The `warning` gives here information about the factors used for scaling of the axis. The plot generated by the `seqdotplot` command is shown in Figure 6.8 (p. 195).

Multiple alignment

Alignment of three or more biological sequences can be done via multiple alignment, which is used for analyses of evolutionary relationships. One of the commands used for this is `multialign`, which uses the progressive method (the hierarchical or tree method) and has the following simplest form:

$$\text{aligned\_seqs = multialign(sequence\_set)}$$

where `aligned_seqs` is a structure or character array output with the aligned sequences; `sequence_set` can be a structure or a character array.

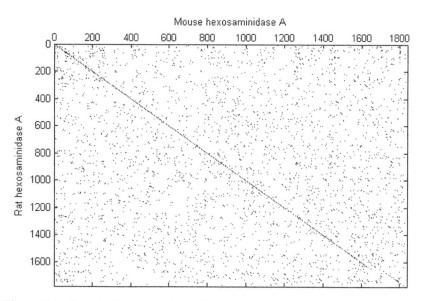

Figure 6.8 Dot plot for mouse and rat hexoaminidase A sequences, generated by the `seqdotplot` command.

Its expanded form may contain the relationships among sequences; by default the command defines distance scores and forms relationships between sequences called the phylogenetic or guide tree.

An example is:

```
>>seqs=strvcat('CGTTAT','TCGTTAC','TAGTTGTGC','GAGTTAATG');
>>ma=multialign(seqs)
ma =
-CGTTAT--
TCGTTA--C
TAGTTGTGC
GAGTTAATG
```

The first command of this example assigns to the `seqs` an array with four sequences containing DNA letters, and the second shows the results of the multiple alignment.

More complicated command forms for scoring matrices, user-given trees, gap-score values and others can be obtained by typing and entering the `help multialign` command.

For better graphical representation of aligned sequences the showalignment(aligned_seq) command is used. For example, an alignment executed by the >>[Score, Alignment] = nwalign('LIFESCIENCE', 'SERIES'); can be represented by typing and entering in the Command Window

>>showalignment(Alignment);

The Aligned Sequences window appears (Figure 6.9):

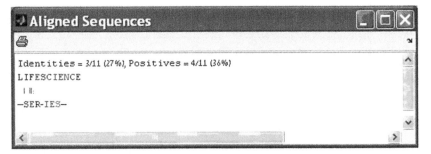

Figure 6.9 Aligned Sequence window generated by the showalignment command; pairwise alignment.

For pairwise alignment the matched and unmatched characters appear respectively in red and magenta.

The graphical representation of multiple alignment (Figure 6.10 shows alignment of four sequences) given in the previous multialign example can be obtained by typing and entering

>>showalignment(ma):

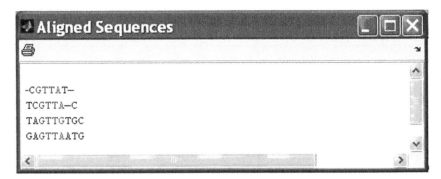

Figure 6.10 Aligned Sequences' window generated by the showalignment command; multiple alignment.

Published by Woodhead Publishing Limited

The columns with highly conserved characters are highlighted and appear in red and conserved characters in magenta.

For long sequences (longer than 64 characters) another form of the `showalignment` command should be used:

```
showalignment(Alignment, 'Columns', Columns_Value)
```

where `Columns_Value` of the `'Columns'` property is the maximal integer number of characters showing in one line; for example, `showalignment(Alignment, 'Columns', 64)` displays up to 64 letters on one line.

Another command that may be used to display multiple alignments is:

```
multialignviewer(alignment)
```

where `alignment` is a structure with the following fields: `Sequence`, which is obligatory, and `Header`, which is optional and is used to display the names of species.

the command opens a viewer that shows the multiple sequence alignment. For example, typing and entering in the Command Window

>> multialignviewer(ma)

the Multiple Sequence Alignment Viewer appears with four aligned sequences (Figure 6.11, p. 197).

The `multialignviewer` command was used in this example with the ma variable, which contains the results of the preceding multiple alignment of four short sequences.

With the Multiple Sequence Alignment Viewer you can also interactively adjust a sequence alignment.

Note: Multiple alignment algorithms do not always lead to optimal results and often need some adjustment. More detailed information on this and other subjects of multiple alignments can be obtained in the appropriate sections of the Bioinformatics toolbox user guide.

6.5 Sequence analysis examples

The examples below show how to apply the commands presented above to some problems in sequence statistics and alignment of paired and multiple sequences.

Published by Woodhead Publishing Limited

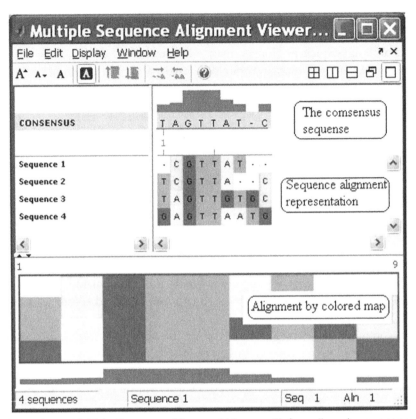

Figure 6.11 Multiple Sequence Alignment Viewer with four aligned sequences generated with the `multialignviewer` command.

6.5.1 Example of sequence statistics

<u>Problem</u>: Retrieve the sequence information for the western gorilla (*Gorilla gorilla*) mitochondrion genome from an online database and determine the mono-, di- and trinucleotide contents with sequence statistics commands.

The first step is to determine the access number for this genome. For this we use the `web` command and explore the NCBI databases. In the Command Window type and enter:

>>web('http://www.ncbi.nlm.nih.gov/')

The NCBI website home page appears in the Web Browser window (Figure 6.1). Then select the 'Genome' option in the first 'Search' line, type `gorilla mitochondrion` into the second Search line and press the 'Search' button. The following window appears.

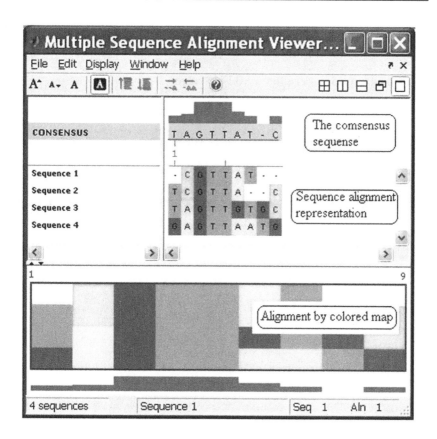

The required accession number is NC_001645, which on clicking provides the information shown at the top of p. 200.

The second step consists of retrieving sequence information from the database to the MATLAB® workspace. For this, type in the Command Window:

```
>> gorilla_mito = getgenbank('NC_001645','sequenceonly',true)
gorilla_mito =
GTTTATGTAGCTTACCTCCCCAAAGCAATACACTGAAAATG
 TTTCGACGGGCTCACATCACCC...
```

The nucleotide sequence from the GenBank database is assigned to the gorilla_mito variable. Information about its size can be obtained by the whos command (see p. 218):

```
>> whos gorilla_mito
   Name              Size              Bytes   Class    Attributes

   Gorilla_mito      1x16364          32728   char
```

Thus the sequence size is 16,364 characters, equivalent to 32,728 bytes.

With the sequence written in the MATLAB® workspace, we can use the following sequence statistics command:

```
>> ntdensity(gorilla_mito)
>> figure, basecount(gorilla_mito,'chart','pie')
ans =
```

A: 5059
C: 5022
G: 2160
T: 4123
>>title('Nucleotide pie chart for a gorilla mitochondrion')
>> figure, dimercount(gorilla_mito,'chart','bar')
ans =
AA: 1586
AC: 1437
AG: 773
AT: 1262
CA: 1481
CC: 1712
CG: 418
CT: 1411
GA: 614
GC: 680
GG: 425
GT: 441
TA: 1378
TC: 1193
TG: 543
TT: 1009
>>title('Dimer bar chart for a gorilla mitochondrion')

>> codoncount(gorilla_mito,'figure',true)

AAA – 158	AAC – 152	AAG – 62	AAT – 127
ACA – 138	ACC – 155	ACG – 37	ACT – 127
AGA – 58	AGC – 83	AGG – 59	AGT – 51
ATA – 112	ATC – 87	ATG – 51	ATT – 93
CAA – 172	CAC – 151	CAG – 73	CAT – 134
CCA – 170	CCC – 199	CCG – 54	CCT – 180
CGA – 34	CGC – 47	CGG – 23	CGT – 32
CTA – 159	CTC – 110	CTG – 67	CTT – 84
GAA – 70	GAC – 48	GAG – 56	GAT – 50
GCA – 64	GCC – 88	GCG – 17	GCT – 63
GGA – 30	GGC – 35	GGG – 25	GGT – 23
GTA – 49	GTC – 29	GTG – 18	GTT – 34
TAA – 164	TAC – 124	TAG – 103	TAT – 143

TCA – 164	TCC – 138	TCG – 53	TCT – 104
TGA – 66	TGC – 37	TGG – 39	TGT – 38
TTA – 113	TTC – 98	TTG – 44	TTT – 88

>>title('Codon count chart for a gorilla mitochondrion')

The first command plots the densities of monomers and combined monomers in the Figure window; two commands of the second command line count the nucleotides, display the values counted and show their distribution on the pie plot in the separate Figure window. The title is generated with the third command line; two commands from the fourth command line count the dimers, display the resulting values and generate the bar chart in the separate Figure window. The title is created here with the fifth command line. The final commands count the codons in the first reading frame, display the resulting values and generate the titled graphical table (heat map). The graphs generated are presented below:

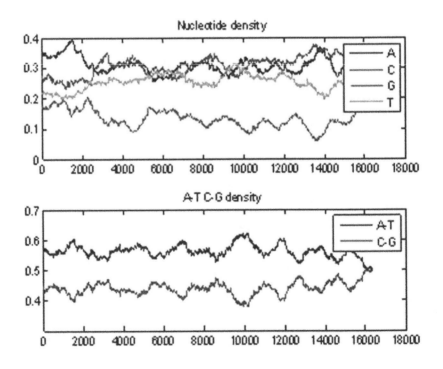

Nucleotide pie chart for a gorilla mitochondrion

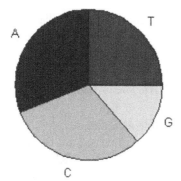

Dimmer bar chart for a gorilla mitochondrion

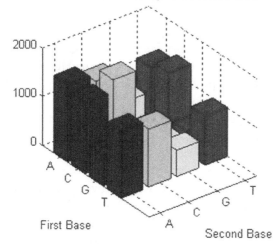

Codon count chart for a gorilla mitochondrion

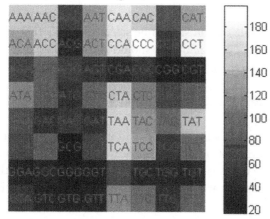

6.5.2 Pairwise alignment example

<u>Problem:</u> Use of primates (chimpanzee, gorilla, etc.) as model organisms for studying human diseases is quite expensive because of the various difficulties involved in acquiring specimens. Hence the need for more accessible alternatives. An example is the search for an organism possessing the enzyme hexosaminidase A (HEXA). The procedure described below assumed that:

- the rat genome is a likely candidate,
- the portions of sequences that should be aligned are the third ORF of the human sequence and the first ORF of the rat sequence.

The problem can be solved via an alignment algorithm in the following steps:

- obtain the accession number for the human gene HEXA;
- obtain the accession number for the rat gene HEXA;
- retrieve the human protein coding sequence from a database and save in the MATLAB® workspace;
- retrieve the rat protein coding sequence from a database and save in the MATLAB® workspace;
- compare the two sequences by an alignment command.

As the first step open the NCBI website by typing and entering in the Command Window:

```
web('http://www.ncbi.nlm.nih.gov/')
```

The NCBI home page appears in the Web Browser window. Now, for example, the Gene option should be selected for the first 'Search' box and the Human hexosaminidase A should be typed into the second box.

Press the 'Search' button, after which the following window appears (p. 205, top):

As can be seen, the accession number for human (*Homo sapiens*) HEXA is NM_000520.

Analogously, to obtain the accession number for the Norway rat, select the Nucleotide option in the first box of the Search engine on the NCBI home page and type Norway rat hexosaminidase A in the second box. After pressing the 'Search' button we have the following window (p. 205, bottom):

Published by Woodhead Publishing Limited

The accession number for brown rat (*Rattus norvegicus*) HEXA is NC_005107.

The next stage is transferring the nucleotide sequences from the GenBank database to the MATLAB® workspace:

```
>> hum_hexa =getgenbank('NC_006482','sequenceonly',true);
>> rat_hexa=getgenbank('NM_001004443','sequenceonly',true);
```

It is now possible to convert the nucleotide sequences into their amino acid counterparts. In accordance with the problem conditions, the third ORF for the human and the first ORF for the rat should be used. The appropriate commands are

```
>> hum_protein = nt2aa(hum_ORF,'frame',3);
>> rat_protein = nt2aa(rat_ORF);
```

For a visual check of similarity, the dot plot can be used (p. 207, top):

```
>>seqdotplot(rat_protein,hum_protein,4,3)
```

The diagonal line on the generated plot shows apparently good alignment of the two sequences.

We can now use the Needleman–Wunsch algorithm for global alignment of the two amino acid sequences:

```
>>[rat_hum_Score, rat_hum_align] = nwalign(rat_protein,hum_protein);
>> for generation of the graphical output
>>showalignment(rat_hum_align)
```

The resultant Aligning Sequence window (p. 207, bottom) shows that the calculated identity of the studied sequences is 47%.

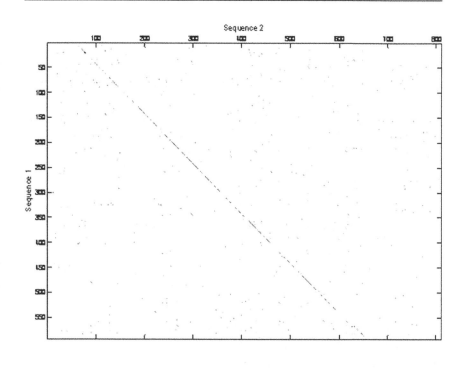

```
Aligned Sequences                                                    [_][□][X]

Identities = 386/815 (47%), Positives = 472/815 (58%)
001 ------------A-C--W--K-G-----S-WPV--G-H------G
            || | : |   : |: || |
001 MPTPGTIRCT*QEPQVQAGSERAGCGSSWGRRRRAASGHVIRR*VTGAPLT*PGSHVASPLREG

013 R----LQALGFAAAGGGVGLLGHGAVALAPVHPNLPPALHPVPQQLPVPVPCRFGRAGGLRC
    |   ||||||:| ||:||  || ||:|:||  ||:|:  ||:  ||: |
065 RPAGHDKLQALVFAAAGGSVRRTGDGPLALASELPNLRPALRPLPEQLSIPVRCQLGRAARLLS

071 PRRGLSTLP*PALRFRLLAPTQLLKKTAAVGEEHSDGLCRHSRM**VS*FGVRRKLHPNH***P
    |||  | |: |||: ||:  :: |||   |||: ||: ||  :|:|||
129 PRRGLPALS*PAFRFRVLAPSLPHRETAYTGEECVGCLCSHTWM*PASYFGVSGELYPDHK**P

135 VFTLL*DCLGRSARSGDFQSACLEVS*GHVLYQQDKDYRLSSIPSPGHTAGYISPLPAVI*HPE
    || |||| | ||| ||:||| | ||:| | ||  : | .. :|||::||
193 VFTPL*DCLGSSPRSGDF*PACLEIC*GHILYQQD*D*GLSPLSSPGLAVGYISPLPATL*HPG

199 HTGCHGVQ*IQRVPLALGGR-LFLPIRELHLPRAHQKGVLQPCYPHLHSTGCEGGH*ICKASGY
    |:||||||:||| | ||: ||:||: ||||| ||||||||||| | ||
257 HSGCHGVQ*IERVPLA-SGR*SFLPI*ELHFSRAHEKGVLQPCHPHLHSTGCEGGH*IRTAPGY

262 PSAGRI*HSWPHFVLGGRCPWIINTLLLWVPSLWHLWTCEPQPQQHLRLHEHILPGDQLCLP*L
    |||:||||| | :: |||| ||| | :  :|||  : ||| :
320 PCACRV*HSWPHFVLGTRYPWITDSLLLWV*ALWHLWTSESOSO*YL*VHEHILLRSOLCLPRF
```
<   >

Published by Woodhead Publishing Limited

6.5.3 Multiple alignment example

<u>Problem</u>: To compare the genetic variability, evolution and other relationships among Hominidae, multiple sequence alignment can be used. The task here is to write and save the script file with the name HuminidaeEx.m that is intended to retrieve, align and show results for four Hominidae mitochondrial DNA (D-loop) sequences – orangutan, chimpanzee, Neanderthal and gorilla. The necessary accession numbers are:

Jari Orangutan – AF451964,
Chimp Troglodytes – AF176766,
Russian Neanderthal – AF254446,
Eastern Lowland Gorilla – AF050738.

The problem will be solved in the following steps:

- Retrieve the sequences from the NCBI database to the Workspace for the mitochondrial DNA (D-loop) of each of the four Hominidae; this can be done with a loop, at each pass of which the accession number in the `getgenbank` command is changed.
- Align the sequences with the `multialign` command.
- Show the results in the Multiple Sequence Alignment Viewer by the `multialignviewer` command.

The script file with the solution of this problem is:

```
% Hominidae alignment
accession_number=['AF451964';'AF176766';'AF254446';'AF176731'];
name=strvcat('Jari Orangutan','Chimp Troglodytes',...
    'Russian Neanderthal','Eastern Lowland Gorilla');
sz=size(accession_number);
for i=1:sz(1)
primate(i).Sequence=getgenbank(accession_number(i,:),'sequenceonly ',
    'true');
primate(i).Header=name(i,:);
end
ma = multialign(primate);
multialignviewer(ma);
```

The sequences and primate names are contained in the `primate` structure in the fields named `Sequence` and `Header`; these names are required by the `multialignviewer` command to display the sequences and their names in the Multiple Sequence Alignment Viewer. After running this file, the window with the four aligned sequences appears:

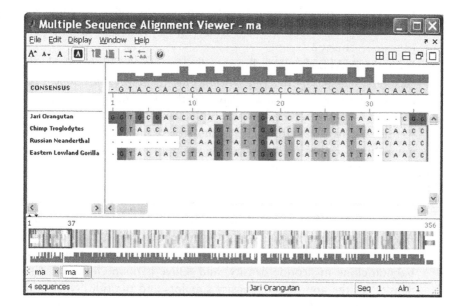

Retrieval of numerous sequences by the `getgenbank` command may be time-consuming and without guaranteed success. It is therefore preferable to save previously retrieved information in a file, and retrieve the sequence data from it for subsequent multiple alignments.

6.6 Questions for self-checking and exercises

1. Which of the given commands may be used for global alignment of two sequences: (a) `swalign`; (b) `nwalign`; (c) `showalignment`; (d) `aa2nt`?
2. The `randseq(20)` command generates: (a) a random sequence containing 20 letters of the amino acid alphabet; (b) a 20-letter RNA random sequence; (c) a 20-letter DNA random sequence?
3. The `dens=ntdensity(seq)` command is intended for: (a) counting amino acid densities in the `seq` sequences and assigning the results to the `dens` structure; (b) counting the density of the A, C, G and T nucleotides in the `seq` and assigning the results to the `A, C, G, T` fields of the `dens` structure?

4. For conversion of an amino acid sequence into a nucleotide sequence the following command is used: (a) `nt2aa`; (b) `aa2nt`; (c) `inv`; (d) `multialign`?

5. For counting codons in a nucleotide sequence the following command should be used: (a) `basecount`; (b) `codonbias`; (c) `dimercount`; (d) `codoncount`?

6. With the MATLAB® Web Browser find in the NCBI GenBank the accession number of the ring-tailed lemur (*Lemur catta*) clone LB2-212N12, retrieve the stored data, save it in the file Lemur.txt, and read the sequence from the appropriate field of the structure stored in this file.

7. Retrieve from the PDB database the stored information about the guinea-pig's oxyhemoglobin – accession number 3AOG; save the data in a file and display the 3D molecular structure. Write all commands in a script file.

8. Get the nucleotide sequence total data set for the olive baboon (*Papio anubis*) clone RP41-187H19 stored in EMBL – accession number AC091381; save it in a file, count the nucleotide densities and bases in the sequence, and display the bases in a pie chart. Add the titles 'Olive baboon clone nucleotide density' and 'Olive baboon clone A-T C-G density' to the appropriate density subplots and the title 'Nucleotide amounts in the Olive baboon clone' to the pie chart. Write all commands in a script file.

9. Write a script file that gets the sequence for mouse insulin receptor (accession number NP_034698 in the GenPept database) and use it as a query sequence in the NCBI BLAST search for alignment of this sequence with those of other species (subject sequences) determined by the BLAST search initiated by a MATLAB® command with 1e-7 E value of the 'expect' property; place the BLAST report into the Workspace and display in the Command Window the names of the species that were compared with the query sequence; these names are stored in the `Hits.Name` field.

10. Write a script file containing the commands that: (a) generate a random 64-letter amino acid sequence, count the amino acids in it, display the results and generate the bar chart; (b) convert the amino acid sequence into nucleotides, count the nucleotides and dimers in the sequence, and generate the bar charts for nucleotides and dimers. Present all three charts in the same plot; add titles for each chart.

11. From the GenPept database get two protein sequences – for yeast and *Escherichia coli*, whose accession numbers are 1S1I_C and 1GLA_G, respectively. Write a script file for global pairwise alignment of these

sequences, display the score and the aligned sequences in the Command Window, and generate a graph with scoring space and winning path.

12. Write a function file for global pairwise alignment of two protein sequences retrieved from GenBank – score and scoring matrix should be displayed; take as input arguments two strings with accession numbers; calculate and display the scoring space and winning path (in a separate window) and score value (in the Command Window) for two residues: Rhesus macaque (*Macaca mulatta*) and human cystic fibrosis transmembrane conductance regulator gene, exon 13; the accession numbers for their nucleotide sequences are AF016937 and M55118, respectively.

13. Solve the preceding problem realizing both global and local pairwise sequence alignment in a single function file; introduce the input argument *N* that allows you to select the desired alignment type: global for *N* = 1 and local for *N* ≠ 1.

14. Write a script file with commands that retrieve (from the GenBank database) nucleotide sequences for the mouse (*Mus musculus*) and human (*Homo sapiens*) hexosaminidase A genes – accession numbers AK080777 and NM_000520, respectively – convert these sequences into the corresponding amino acid sequences, execute global pairwise alignment, display the scoring space and winning path in a separate window, and display the score value in the Command Window.

15. Get nucleotide sequences from the GenBank database for the cystic fibrosis transmembrane conductance regulator (CFTR) of:
bull (*Bos taurus*) – accession number NM_174018,
human (*Homo sapiens*) – accession number U20418,
mouse (*Mus musculus*) – accession number NM_021050.

Convert nucleotide into protein sequences; execute multiple alignment and present the results in the Alignment Sequences window. Write all commands in a script file.

6.7 Answers to selected exercises

2. (c) a 20-letter DNA random sequence.
7.
```
>> Ch5_7
              Header:  [1x1 struct]
               Title:  [2x57 char]
```

```
       Compound:  [8x46 char]
         Source:  [8x37 char]
       Keywords:  [2x60 char]
 ExperimentData:  'X-RAY DIFFRACTION'
        Authors:
'S.ETTI,G.SHANMUGAM,P.KARTHE,K.GUNASEKARAN'
   RevisionDate:  [1x1 struct]
        Journal:  [1x1 struct]
        Remark2:  [1x1 struct]
        Remark3:  [1x1 struct]
        Remark4:  [2x59 char]
      Remark100:  [3x59 char]
      Remark200:  [49x59 char]
      Remark280:  [7x59 char]
      Remark290:  [48x59 char]
      Remark300:  [6x59 char]
      Remark350:  [27x59 char]
      Remark465:  [11x59 char]
      Remark470:  [7x59 char]
      Remark500:  [100x59 char]
      Remark620:  [28x59 char]
      Remark800:  [17x59 char]
      Remark900:  [7x59 char]
   DBReferences:  [1x2 struct]
SequenceConflicts:  [1x3 struct]
       Sequence:  [1x2 struct]
      Heterogen:  [1x4 struct]
  HeterogenName:  [1x2 struct]
HeterogenSynonym:  [1x1 struct]
        Formula:  [1x3 struct]
          Helix:  [1x18 struct]
           Link:  [1x6 struct]
           Site:  [1x1 struct]
         Cryst1:  [1x1 struct]
        OriginX:  [1x3 struct]
          Scale:  [1x3 struct]
          Model:  [1x1 struct]
   Connectivity:  [1x94 struct]
         Master:  [1x1 struct]
      SearchURL:  3A0G
```

Published by Woodhead Publishing Limited

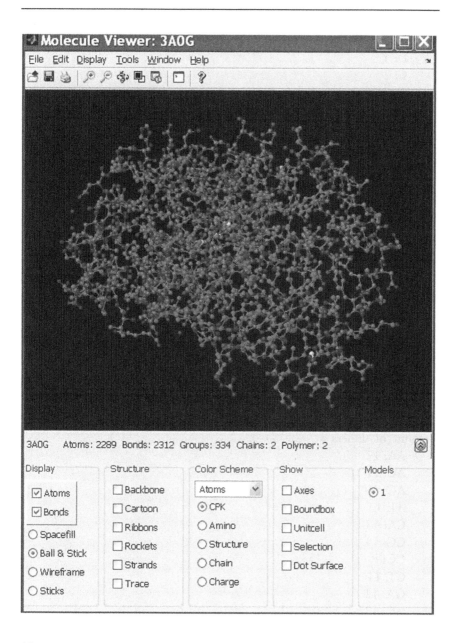

10.
>> Ch5_10
amount_of_aminoacids =
   A: 4
   R: 0

Published by Woodhead Publishing Limited

```
     N: 3
     D: 1
     C: 2
     Q: 2
     E: 5
     G: 6
     H: 2
     I: 3
     L: 2
     K: 4
     M: 4
     F: 5
     P: 6
     S: 1
     T: 6
     W: 2
     Y: 1
     V: 5
amount_of_nucleotides =
     A: 51
     C: 47
     G: 49
     T: 45
amount_of_dimers =
     AA: 17
     AC: 11
     AG: 10
     AT: 13
     CA: 13
     CC: 14
     CG: 9
     CT: 11
     GA: 12
     GC: 12
     GG: 14
     GT: 10
     TA: 9
     TC: 10
     TG: 15
     TT: 11
```

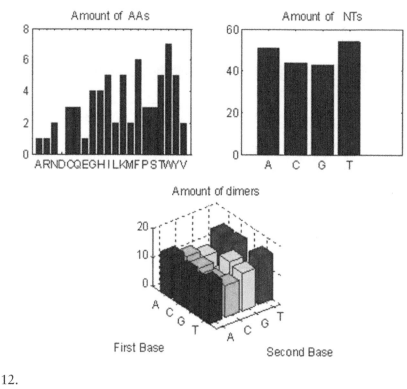

12.

```
>> score=Ch5_12('AF016937','M55118',2)
Score =
-246
```

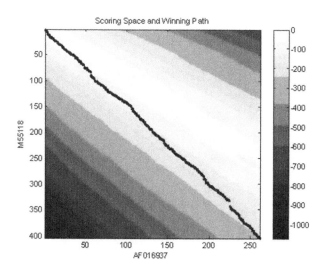

15.

```
>> Ch5_15
```

Note

1. Other grading modes are possible, for example 2 for matches and −1 for mismatches/inserts.

Appendix: MATLAB® characters, operators and commands

Characters, arithmetic, rational and logical operators

Character, operator	Description	Page(s)
+	Addition	6, 19
–	Subtraction	6
*	Scalar and matrix multiplication	6, 20
.*	Element-wise multiplication	28
/	Right division	6, 26
\	Left division	6, 26
./	Element-wise right division	28
.\	Element-wise left division	28
^	Exponentiation	6
.^	Element-wise exponentiation	28
=	Assignment	6
%	Percent; use for comments and for output format specification	6, 11
( )	Parentheses; use for input arguments and in matrix addressing	6, 15, 18
[ ]	Brackets; use for vector, matrix, array elements input	15, 17
(space)	Space; separates elements into arrays, and adds into output specifications	15, 18
,	Comma; separates elements into arrays, and commands on the same line	6, 15
:	Semicolon	6
...	Ellipsis	6
<	Less than	34
>	Greater than	34
<=	Less than or equal to	34
>=	Greater than or equal to	34
= =	Equal	34
~=	Not equal	34
&	Logical AND	35
		Logical OR	35
~	Logical NOT	35

Published by Woodhead Publishing Limited

Help and managing commands

Command	Description	Page
clear	Remove variables from the Workspace	10
clc	Clear the Command Window	6
close	Closes one or more Figure Windows	53
help	Displays explanations for commands	7
help graph2d	Displays list of 2D graphs	76
help graph3d	Displays list of 3D graphs	76
help specgraph	Displays list of specialized graph commands	76
doc	Displays HTML documentation in the Help window	9
function	Creates a new function	100
global	Declares a global variable	102
lookfor	Search for the word in all help entries	9
ver	Displays versions of the MATLAB® products	171
who	Displays variables stored in the Workspace	10
whos	Displays Workspace variables and additional information about the variables	10

Predefined variables and elementary math functions

Variable, Function	Description	Page
abs	Absolute value	7
ans	Last calculated or defined value	8
exp	Exponential	7
factorial	Factorial function	7
floor	Round off toward minus infinity	7
i	$\sqrt{-1}$	10
inf	Infinity	10
j	The same as i	10
log	Natural logarithm	7
log10	Decimal logarithm	7
NaN	Not a number	10
pi	Number π	7
round	Round off toward nearest integer	7
sqrt	Square root	7
sin	Sine	7
cos	Cosine	7
tan	Tangent	7
cot	Cotangent	7
asin	Inverse sine	7
acos	Inverse cosine	7
atan	Inverse tangent	7
acot	Inverse cotangent	7

Published by Woodhead Publishing Limited

Relational and logical commands

Character, command	Description	Page
and	Logical AND	35
or	Logical OR	35
not	Logical NOT	35
find	Finds indices of certain elements of array	35

Flow control commands

Command	Description	Page(s)
else	Is used with if	37
elseif	conditionally execute if statement condition	37
end	Terminates scope of for, while, if statements, or serves as last index	17, 37
for	Repeat execution of command/s	37
if	Conditionally execute	37
while	Repeat execution of command/s	38

Array, matrix, and vector commands

Command	Description	Page(s)
colon (:)	Is used for creating a vector	15
det	Calculates a determinant	27
diag	Creates a diagonal matrix from a vector	26
eye	Creates a unit matrix	22, 26
inv	Calculates the inverse matrix	22
length	Number of elements in vector	26
reshape	Changes size of a matrix	26
linspace	Generates a linearly spaced vector	16
max	Returns maximal value	27
min	Returns minimal value	27
mean	Calculates mean value	27
ones	Creates an array with ones	25
rand	Generates an array with uniformly distributed random numbers	25
randi	Generates an array with integer random numbers from uniform discrete distribution	25, 27
randn	Generates an array with normally distributed numbers	25
size	Size of array/matrix	26
sort	Arranges elements in ascending or descending order	27

Published by Woodhead Publishing Limited

(Continued)

Command	Description	Page
std	Calculates standard deviation	27
strvcat	Concatenates strings vertically	26
sum	Calculates sum of elements	27
transpose (')	Transposes elements of an array	19
zeros	Creates an array with zeros	25

Input, output, display format, and convert commands

Command	Description	Page
disp	Displays output	11
fprintf	Displays or saves formatted output	11
input	Prompts to user input	99
format	Sets current output format	8
num2str	Converts numbers to a string	27

Two- and three-dimensional plotting and plot formatting

Command	Description	Page
axis	Controls axis scaling and appearance	57
bar	Generates vertical bars on the plot	78
bar3	Generates 3D vertical bars on the plot	78
box, box on/off	Adds a box to the current axes, keeps/removes a box on the axes	68
clabel	Labels iso-level lines	78
colormap	Sets colors	68
contour	Creates a 2D-contour plot	78
contour3	Creates a 3D-contour plot	78
cylinder	Generates a cylinder	77
errorbar	Creates a plot with error-bounded points	73
figure	Creates the Figure window	77
fplot	Creates a 2D plot of a function	77
gtext	Adds text with the help of the mouse	59
grid	Adds grid lines	57
hist	Plots a histogram	73
hold on, hold of	Keeps current graph open, ends hold on	53
legend	Adds a legend to the plot	59
loglog	Generates a 2D plot with log axes	77
mesh	Creates a 3D plot with meshed surface	66
meshgrid	Creates X,Y matrices for further plotting	65

(Continued)

Command	Description	Page
pie	Creates a 2D pie plot	79
pie3	Creates a 3D pie plot	79
plot	Creates a 2D plot	53
plot3	Creates a 3D plot with points and/or lines	63
polar	Creates a 2D plot in polar coordinates	77
polyfit	Fits the data by a polynomial	114
polyval	Evaluates the polynomial value	114
rotate3d	Interactively rotates a 3D plot	72
semilogx	Creates a 2D plot with log-scaled x-axis	75
semilogy	Creates a 2D plot with log-scaled y-axis	75
sphere	Generates a sphere plot	77
stem	Creates a 2D stem plot	79
subplot	Places multiple plots on the same page	55
surf	Creates a 3D surface plot	66
surfc	Generates surface and counter plots together	78
text	Adds text to the plot	59
title	Adds a caption to the plot	59
view	Specifies a viewpoint for 3D graph	70
xlabel	Adds a label to x-axis	59
ylabel	Adds a label to y-axis	59
zlabel	Adds a label to z-axis	63

Math functions, integration and differentiation

Command	Description	Page
diff	Calculates a difference, approximates a derivative	110
fzero	Solves a one-variable equation	107
interp1	One-dimensional interpolation	105
quad	Numerical integration with Simpson's rule	108
trapz	Numerical integration with the trapezoidal rule	109

Ordinary and partial equation solvers

Command	Description	Page
ode15s	Solves stiff ODEs	135
ode15i	Solves implicit ODEs	135
ode23	Solves non-stiff ODEs	135
ode23s	Solves stiff ODEs	135
ode23t	Solves stiff ODEs	135
ode23tb	Solves stiff ODEs	135
ode45	Solves non-stiff ODEs	135
ode113	Solves non-stiff ODEs	135
odeset	Sets ODE options	170
pdepe	Solves 1D parabolic and elliptic PDEs	153

Published by Woodhead Publishing Limited

Bank access and sequence analysis

Command	Description	Page
aa2nt	Converts a sequence of amino acids to a nucleotide sequence	183
aacount	Counts amino acids in a sequence	184
aminolookup	Finds amino acid or codon codes	188
basecount	Counts nucleotides in sequence	188
baselookup	Returns nucleotide codes	188
blastncbi	Requests NCBI BLAST report and its ID	181
codoncount	Calculates codons in nucleotide sequence	189
dimercount	Calculates dimers in nucleotide sequence	189
emblread	Reads data stored in EMBL-formatted file	181
genbankread	Reads data stored in GenBank-formatted file	178
genpeptread	Reads data stored in GenPept-formatted file	180
getblast	Gets BLAST report generated in NCBI website	182
getgenbank	Gets data from GenBank database	178
getgenpept	Gets sequence info from GenPept database	179
getgeodata	Gets data in GEO format	180
getembl	Gets sequence info from EMBL database	179
getpdb	Gets protein 3D data from PDB database	180
molviewer	Generates 3D molecule image	181
molweight	Calculates molecular weight of amino acids	189
multialign	Aligns multiple sequences	194
multialignviewer	Displays and interactively adjusts multiple sequence alignment	197
nt2aa	Converts nucleotide to amino acid sequence	206
ntdensity	Plots nucleotide densities along sequence	186
nwalign	Aligns two sequences globally	192
pdbread	Reads data stored in PDB-formatted file	181
randseq	Generates letter-coded random sequence	183
seqdotplot	Generates dot plot for two sequences	194
seqshoworfs	Shows open reading frames	184
showalignment	Displays results of alignment	196
swalign	Aligns two sequences locally	209
web	Opens website or file in Web Browser window	175

Published by Woodhead Publishing Limited

Index

CPSIA information can be obtained at www.ICGtesting.com
Printed in the USA
LVOW05*1125261114

415562LV00001B/100/P

9 781907 568046